THE
BEAUTY
of DETOURS

THE BEAUTY *of* DETOURS

A BATESONIAN PHILOSOPHY OF TECHNOLOGY

Yoni Van Den Eede

Cover art: "Chemicals Are Ready, Prepare to Flash" by Ellen Vandeperre (Vep), © Ellen Vandeperre.

Published by State University of New York Press, Albany

© 2019 State University of New York Press

All rights reserved

No part of this book may be used or reproduced in any manner whatsoever without written permission. No part of this book may be stored in a retrieval system or transmitted in any form or by any means including electronic, electrostatic, magnetic tape, mechanical, photocopying, recording, or otherwise without the prior permission in writing of the publisher.

For information, contact State University of New York Press, Albany, NY
www.sunypress.edu

Library of Congress Cataloging-in-Publication Data

Names: Eede, Yoni van den, author.
Title: The beauty of detours : a Batesonian philosophy of technology / Yoni Van Den Eede.
Description: Albany : State University of New York, 2019. | Includes bibliographical references and index.
Identifiers: LCCN 2019011364 | ISBN 9781438477114 (hardcover : alk. paper) | ISBN 9781438477121 (pbk. : alk. paper) | ISBN 9781438477138 (ebook)
Subjects: LCSH: Technology—Philosophy. | Bateson, Gregory, 1904–1980.
Classification: LCC T14 .E33 2019 | DDC 601—dc23
LC record available at https://lccn.loc.gov/2019011364

10 9 8 7 6 5 4 3 2 1

Contents

Preface — vii

Acknowledgments — xiii

Introduction — xvii

Part I: Laying the Groundwork

Chapter 1 Gregory Bateson's "Ecology of Mind" — 3

Chapter 2 Philosophy of Technology — 17

Part II: Bateson and Technology

Chapter 3 Between One and Two: Epistemology/Ontology — 43

Chapter 4 Conscious Purpose — 59

Chapter 5 Remediating Conscious Purpose — 79

Part III: The Art of Living with Technology

Chapter 6 Toward Batesonian Philosophy of Technology — 99

Chapter 7 The Art of Living with Technology — 131

Notes	177
References	201
Index	215

Preface

"We are not outside the ecology for which we plan—we are always and inevitably a part of it."[1] Gregory Bateson wrote these words in 1970. He passed away ten years later—before he could see the rise of the Reagan-Thatcher era, the full deployment of neoliberalism, the digital revolution, and the increasingly clear effects of climate change. But he saw the writing on the wall. He noticed the tide approaching, right before the wave rolled over us. As a pioneer of ecological thought, he was concerned about environmental degradation and the exploitation of nature. He warned about overly economic reasoning. Worried that we would be reducing reality to something it is not (or at least not exclusively), he pointed to the risks of single-sided quantification and calculation. Overall, his prime concern was with dynamics in which, due to thinking patterns becoming ingrained and turning rigid, and thus acquiring a taken-for-granted stature, some species, group, organism, or individual—*any* species, group, organism, or individual—comes to suffer at the expense of another.

Those developments that Bateson didn't get to see have set us on the way to where we are now, that is, the post-truth world of Donald Trump, in which unashamed lying, bigotry, racism, and sexism have once again become "normal." It is also the ecological limbo we are in, where we are increasingly aware of the environment's terrible condition but cannot seem to move ourselves to act effectively upon that insight. And it is the invisible algorithmic architecture that's starting to envelop most of our everyday lives, with our daily occupations becoming more and more imperceptibly organized, shaped, and steered by algorithms.

I like to wonder what Bateson would have thought about our times. How would he have looked upon "alternative facts," for one? I imagine him having let out his typical chuckle. But as often in his case, that chuckle would have been coupled with grave concern. The chuckle represents the recognition: ah yes, this is humankind—in all its ridiculous, painful, beautiful, horrible glory. But superposed onto that comes the critical concern: it shouldn't have to be this way; *it could be otherwise*. Chuckle and concern are at the same time there, in equal measures. Things are amusing, *and* they are serious.

Is there still such a thing as humankind? After decades of posthumanism, postmodernism, poststructuralism, and the like, we are inclined to say: no. The human being seamlessly flows over into "other things": nature, technology, society. In Foucault's famous words, the human being is "a face drawn in sand."[2] Bateson came, from another angle, upon a similar idea: "We are always and inevitably a part of" the ecology. We should learn to see and define ourselves that way: as network structures, essentially interlocked with other species, organisms, the biosphere in whole. Yet it would be a mistake—a moral mistake—to leave it at that, shrug our shoulders in resignation and say: if we're nothing special, if we're just "part of," it doesn't really matter anymore what "we" do. It's all one ecology anyway; what should it matter? Well, despite that hybrid, dispersed, spread-out nature, *we* still have responsibility. This is one of the main insights that the currently much discussed notion of the Anthropocene attests to. No matter how authors have questioned the *anthropos* part in that term—and proposed all sorts of alternative *cenes* (Chthulucene, Capitalocene, et cetera)[3]—there always remains the underlying recognition of *us* having bit off more than we can chew. Humans are the prime cause of climate change, not dolphins, the Amazon, or bacteria. So we face a burden, a challenge either to deal with the consequences thereof or to correct for our mistakes (and probably a mix of both). In this sense, *we* are responsible. And in this sense, at minimum, there still exists a "we." There remains at least one possible definition of humankind, and that definition is ethical.

But because we have some ethical responsibility—we face the consequences of what we've done—there must also have been a part of us inclining in the first place toward those harmful acts. Reverse-engineering the human being, starting from the observation of the result of its deeds—a wrecked environment, cynicism, power abuse—and tracing back from there—what caused this?—we must eventually bump into traits of the human animal that *made us this way*. Bateson's work is all about this. What makes systems go haywire? What makes ecologies go out of balance? The human being, or what's left of it, seems to harbor in itself a quality that makes it badly attuned to ecological balances. In Bateson's view, that is our ability to *chase after our purposes*—to achieve the goals we set out to achieve. Strangely, as such, that quality is a good thing. It is a trait we share with other living beings: the cat goes out to catch a mouse; the buffalo moves to the pool to drink; the tree stretches out its roots to absorb nutrients. However, in humans, this goal-orientedness has throughout history taken on globe-covering dimensions. We built pyramids, invented capitalism, installed the Internet. Often with good intentions, we set after a purpose, and got to work. Need to grow more crops to feed people? Let's develop fertilizer. A couple of decades later, we find fertilizer causing major environmental harm and health problems. Or: let's put people in touch with each other. Here's a solution: social media. Some years later, it turns out that social media are a perfect instrument for illicitly influencing people's political opinions. Somewhere along the line in many of our goal-achieving processes, a threshold gets reached, and the system starts to go out of control. What was eventually meant to heighten control—a good thing—spins out into wider territory somehow and we lose control—a bad thing; or at the very least an ambiguous thing.

What lies at the base of this bizarre kind of dialectic of Enlightenment? It is the nature-wide drive to get what one wants. But for some reason—intelligence? social skills? speech? technology?—humans seem to have been, and are still, most successful in spreading out their blanket of goal achievement across the earth. In this sense, our times probably do not differ so much from those in which Bateson lived. We may have that idea: that we live in exceptional times—think, for instance, Trump again—but that is probably because our horizon of a human life is too short to truly oversee historical currents and constants. The dynamics that Bateson sought to describe can still be recognized in us. Many of us still think—again, notwithstanding postmodernism, posthumanism, and so on—that we humans are the pinnacle, if not of creation, then certainly of evolution. If we don't think it explicitly, we must do so implicitly, because this is simply how more often than not we appear to act upon this earth: strutting around, knocking things over, unashamedly, hardly looking back, hardly looking forward.

So what happens in between a well-meant intention and a nefarious end result? Production becomes pollution. Rationalization turns into suffocation. Calculation morphs into alienation. Crosscutting all of this, Bateson insists, is a very simple idea, really: trying to *get what you want* is never done in isolation. You may in the process not have much regard for anything else—but the world is still there. You may only have wanted to throw the stone, but your actions create ripples on the pool of life. Given that we lead our lives wrapped up, entangled in constellations in which everything we do is connected to something else (and that to still something else, and so on . . . indefinitely), there can never be something like a pure neutral act. We may like to believe so. In fact, we constantly do. We go around as individuals, organizations, companies, governments, and societies assuming that the world is made up of linear, mathematical, A-to-B sequences. "If I want to get from point A to point B, and this is the shortest route, all I have to do is simply follow it." While actually this route crisscrosses others. Or, paving your road, you may be ravaging other roads. There is no road apart from any other. Even "walking the lone road" can only ever be done in a world that is already there—with routes to stray from, stay on, neglect, demolish, expand, build anew. Once again, as the quote at the beginning reads, "We are always and inevitably a part of" the ecology.

Yet, lest one is inclined to put the emphasis on "ecology" alone in the phrase, there is indubitably also indeed mention of our human disposition and need to *plan*. Even in our attempt to create a *healthy* ecology, we cannot but plan (plan *for* our ecology, as Bateson says), and this planning must necessarily be part and parcel of the ecology we're in. That means: with everything we do to deliberately move in a certain direction, no matter how "right" or legitimate that direction is, we are bound to generate unforeseen side effects. We cannot escape the loop. We are the ouroboros, biting our own tail.

Nowhere more does this idea get instantiated than in technology. Like nothing else, technology represents and embodies our wanting to achieve something, with an eye on improvement, on making things better. Technology is humankind's prime way of facilitating its purpose-seeking endeavors. However, not without the catch: any technology, the "materialized" act of planning, will frustrate our best laid plans. It may not act as planned, or it may help to produce unplanned by-products. To mitigate the harm done, then, we mount against it *new* planning endeavors. But these will inescapably replicate the pattern: engender new by-products. It's a paradox. In order to remedy the problem, we must make new problems.

But it seems the only way. Banning technology is not an option—we *are* technology. This is a central insight in contemporary philosophy of technology: on a deep level, humans are indistinguishable from technology. Devising, designing, and deploying technology comes natural to us; we *have to*. That nevertheless does not preclude the aforementioned issues: the side effects, the by-products. Despite our fundamental technological character, we still have to cope with those. Moreover, today, something else is beginning to severely complicate things: technology becomes harder and harder to "see."

There is a hint of paradox here as well. Dramatic technological developments such as nanotechnology, biotechnology, and artificial intelligence *conspicuously* thin the boundary between the human being and technology. Where does the nanotech implant end and the human begin? How should "my" thinking process be differentiated from the machine learning algorithms helping me to decide on a purchase? It will become increasingly difficult to answer these kinds of questions. In all its dramatic conspicuousness, the thinning boundary makes technology *less visible*. Technology is starting to leave the radar. It's becoming tougher and tougher to tell where it is, and what it is. So how then should we make sense of it? How should we delineate and define it? And how should we continue to get a grip on the trajectory from good intention to bad (or at least ambiguous) end result?

This book is about those questions, and it frames them from the perspective of Bateson's work. At the same time, Bateson's work is placed in an unusual setting: it is resituated within the philosophy of technology. The philosophy of technology itself struggles to grasp disappearing technology—it tries to catch a glimpse of technology, just as it disappears. Technology *is* this disappearing, in a way, but we still *have a need* for some well-defined notion of it. This only seems like an impossible contradistinction. We must learn to "think double," to see double: simultaneously this *and* that. Things are *here* and *there*. But it helps in that exercise if we can arrive at matters from a fresh angle. In this book, that new angle is *purpose*. However, purpose itself is also looked at from a fresh perspective—to a certain extent thinking with Bateson against Bateson. Purpose is no exclusively linear business. It can be simple and complex at the same time, and the idea is not to pick sides, but to trace the course that leads from here to there. Bateson's work is about bridging gaps while leaving the gaps intact. In fact, the gap becomes a constitutive element. We humans are striving beings. We picture ourselves a target and then storm toward it, like madmen. All the while, we *also* are network-like structures, fluid, flowing over into many other things. So our linear going-from-A-to-B must eternally be frustrated, as we collide with countless other A-to-Bs. Purpose comes to clash with purpose. Our narrow purposive endeavor clashes with wider purposive constellations. But it's not an issue of having to choose between those two sides. We are both-in-one and should put our energies into coming to grips with this situation rather than finding a "pure" resolution, thinking that the world consists in either one of them. We are the gap.

Learning to think double in this way, to get used to multiplicity (instead of craving for "one"), is a condition for artfully living with technology—for artfully living as such. Bateson knew this. And in opening ourselves up to multiplicity lies the liberation from asphyxiating despair. This is where we combine the chuckle with the concern. As Marshall McLuhan stated beautifully, "The future masters of technology will have to be lighthearted and intelligent. The machine easily masters the grim and the dumb."[4] Crucial is the "and": we need to be lighthearted *and* intelligent. It's about that combination. It won't do to be grim and intelligent, or lighthearted and dumb—those combinations will only lead to despair or to resignation. No, what we require is a cheerful acceptance of the situation as it shows itself to us, and then a sensible, clever response to that. The idea is to first try to understand things as best as possible, and on the basis of that understanding choose a path that's as wise as possible. There is merit in trying to lay bare how the world really is, and how we humans really are—and then look upon those findings with kindness, with kind amusement; like one would lovingly watch a cat scratch one's favorite furniture. We realize the animal does not *mean* it that way—this is just how it is; this is just what happens. But we may try to train the cat, still lovingly, to act differently (or—certainly also a worthy option—find peace in the thought that always-clean, mint-condition furniture isn't all that important). We let out a chuckle *and* formulate concern. The chuckle without concern would be merely evasive giddiness. Concern without the chuckle would be sour, just as indolent, criticism. What gets us to our feet is the realization that we are part of something, of a framework, *and* that we have a voice in that framework, a capacity to change things—because it *is us*.

We are part of the ecology for which we plan. We humans are nothing special, but at the same time, perhaps we are. Perhaps we are something admirable; yet not because we are so powerful and smart. Rather, we can pat ourselves on the shoulders for our potentiality: our ability to *become* better at understanding ourselves and our surroundings, and to treat both more respectfully and justly than we previously did. However, that is a never-finished task, always temporary, always provisional—a step-by-step trying out of things, failing, trying again, failing again. The point is to become more aware of how well-envisioned purposes spin out into chaos.

Acknowledgments

One of the best reading suggestions I ever received came from Mary Catherine Bateson. At the 2010 Annual Convention of the Media Ecology Association, I coincidentally wound up with her and her sister Nora Bateson at the breakfast table. We started talking. They were so kind and generous. The previous night, Nora had premiered her documentary *An Ecology of Mind: A Daughter's Portrait of Gregory Bateson*, and it had floored me. In watching it, a seed had been planted. I had only had a vague idea of what Gregory Bateson stood for. In my work on Marshall McLuhan and media ecology I had bumped into his work now and then, and I had sampled some of *Steps to an Ecology of Mind*—but it hadn't made much sense to me. I couldn't connect the dots. At the breakfast table, I naively confessed to this. Instead of brushing me off or belittling me for my ignorance—those kinds of reaction, although generally rare, are always possible in academic dialogue!—Mary Catherine acknowledged that my befuddlement was not uncommon, and suggested that a much better place to start than *Steps* was the later *Mind and Nature*. *Mind and Nature* offers the beginning of a synthesizing outlook that *Steps* does not; so this was a more natural entry point. Soon after that, I ordered a copy and started reading it, and there indeed it was: things began to gel in my mind. Mary Catherine had been right on target. Eventually, Gregory Bateson's work would become a force of major significance in my life and thinking. And what started at that breakfast table, and the night before with the documentary screening, finds a culmination in this book. I am grateful to Mary Catherine and Nora Bateson for patiently hearing me out on that morning, and for gracefully advising and inspiring me.

This was a germinating event, but along the way many people have helped me in one way or another with the writing of the book. I can only name those who did so most explicitly, but I extend a general thank you to the communities around the Society for Phenomenology and Media, the Society for Philosophy and Technology, the Media Ecology Association and the Society for Social Studies of Science (4S). Ideas developed in this book were often tried out at those venues, and many conversations, comments, and suggestions along the way helped to give shape to the final product. Also, the band of postphenomenologists that gathers regularly at some of these conferences deserves a warm mention, for offering an environment in which thoughts can come to fruition in an atmosphere of constructive collegiality and friendship. This always has been in the first instance the merit of postphenomenology's helmsman Don Ihde and his untiring generosity and encouragement (extended graciously even if he doesn't agree with you—as is regularly the case with me), and I thank Don specifically for his guidance and support. Also Robert C. Scharff's and Mark Coeckelbergh's words of encouragement and advice were instrumental in an early phase of the project. Dominic Smith and Galit Wellner invited me to speak at events about ideas presented in the book, and those moments were crucial for it. My gratitude goes out to them, and for their support in general.

Thank you also for inspiration, support, input, or comments at several stages of the manuscript: Gert Goeminne, Pieter Lemmens, Natasha Dow Schüll, Marc Van den Bossche, Hub Zwart. A special mention is for Richard Lewis, who gave me crucial suggestions with regard to structuring the material, and diligently proofread chunks of text. Thank you, Ellen Vandeperre, for kindly giving me permission to use the cover image. To the anonymous reviewers I am grateful for their profound scrutiny and helpful feedback. I am thankful to acquisitions editor Andrew Kenyon and the people at SUNY Press for taking this project on and expertly guiding it to its completion. Also, I owe a particular thanks to Nora Bateson and Phillip Guddemi of the Bateson Idea Group for providing me with unpublished material by Gregory Bateson and granting me permission to cite it.

Finally, as ever, my most fundamental thanks go to An, for putting up with my inevitable writing mood swings, but most importantly for always nudging me gently back to tenacity.

Portions of this manuscript were previously published in an earlier version and are integrated here, in updated and/or reworked form, in the bigger framework of the book.

The introduction, chapters 1, 2, 4, and 5 incorporate excerpts from "The (Im)Possible Grasp of Networked Realities: Disclosing Gregory Bateson's Work for the Study of Technology," *Human Studies* 39, no. 4

(2016): 601–20. The introduction also incorporates an excerpt from Yoni Van Den Eede, Gert Goeminne, and Marc Van den Bossche, "The Art of Living with Technology: Turning over Philosophy of Technology's Empirical Turn," *Foundations of Science* 22, no. 2 (2017): 235–46.

Chapter 2 reproduces excerpts from "Where Is the Human? Beyond the Enhancement Debate," *Science, Technology, & Human Values* 40, no. 1 (2015): 149–62, and from (my contribution to) Gert Goeminne, Tamar Sharon, Yoni Van Den Eede, Bregham Dalgliesh, and Michel Puech, "Book Symposium on *Homo Sapiens Technologicus: Philosophie de la Technologie Contemporaine, Philosophie de la Sagesse Contemporaine*," *Philosophy & Technology* 27, no. 4 (2014): 581–608.

Chapters 3, 6, and 7 incorporate extracts from "Concrete/Abstract: Sketches for a Self-Reflexive Epistemology of Technology Use," *Foundations of Science* 22, no. 2 (2017): 433–42.

Chapter 4 reproduces extracts from "The Sense of Purpose: A Probe, by Way of 'Old' and 'New' TV Sci-Fi," *Proceedings of The Society for Phenomenology and Media* 3, San Diego: The Society for Phenomenology and Media (2016): 105–13, and from "Media Use, Purpose, and Autopoiesis," *Glimpse: Phenomenology and Media* 16 (2015): 77–81.

Chapter 6 contains excerpts from the book chapter "The Mediumness of World: A Love Triangle of Postphenomenology, Media Ecology, and Object-Oriented Philosophy," in *Postphenomenology and Media: Essays on Human–Media–World Relations*, edited by Yoni Van Den Eede, Stacey O'Neal Irwin, and Galit Wellner, 229–50 (Lanham, MD: Lexington Books, 2017) and from "Variations upon Ihde's *Husserl's Missing Technologies*," *Techné: Research in Philosophy and Technology* 20, no. 2 (2016): 105–11.

Chapter 7 contains an extract from "Beyond the Concrete: Toward an Art of Living with Abstract Conditions," *Foundations of Science* 22, no. 2 (2017): 451–54.

I thank the following publishers and organizations for permission to reproduce this material: Philosophy Documentation Center, Rowman & Littlefield, SAGE, Society for Phenomenology and Media, and Springer.

Introduction

Take a minute to think about *technology*. There's a good chance you will be conjuring up almost instinctively before your mind's eye images of fiberglass cables, computers, switchboards, pipelines, factories, big and small machines. Nowadays you may also throw in touchscreens, colorful app interfaces, and more abstract notions such as algorithms. You might even contemplate visions of cyborgs, by now a well-worn staple of science-fiction movies.

Nothing could be more natural. We spontaneously imagine a range of devices—things—when prodded to give a definition of technology, be they material or immaterial, hardware or software. Ingrained in our worldview lies a well-established assumption of what technology is: something we can look at, point at. Something we at least can *name*. "We" are standing on this side, "technology" is over on the other. This common-sense definition seems harmless enough. However, technological developments are completely outmoding it.

Technology is oozing into everything. Think of—and this can only be a minimal sample of all possible examples—neural implants that control moods, 3D-printing of living tissue and organs, contact lenses recording video or enabling night vision, augmented reality apps of all sorts, Google DeepMind algorithms beating the champion of the game of Go, genome editing, and more. Every week brings new technological, scientific, and medical breakthroughs. More and more domains previously untouched by technology become "technologically mediated."

We may look back on the recent past and wonder about how much more is *possible* nowadays. Diseases that wrecked whole populations not so long ago—polio, tuberculosis—have been all but eradicated or can be treated. The reader born before 1990 remembers the effort it took to communicate with people far away. Now you drop a simple e-mail and in a split second, the person in the other corner of the globe can read what you've written. I keep finding it endlessly intriguing how we have gone in just a good twenty years from a world in which information was hard to come by—requiring a visit to the library, for example—to our current world in which knowledge is at the tip of our fingers, just a few mouse clicks away. We have means to extract energy from sunlight, from wind. There are robot vacuum cleaners sniffing out our houses. We generally—that is, in the West—live long, comfortable, and relatively painless lives.

Now try to look forward into the future. Extrapolating from current developments, how many more possibilities can we imagine? Who knows: in 2050, maybe we will be able to cure cancer, or it will not even exist at all anymore. People will have neural implants that enable direct access to the Internet.[1] Forget the mouse clicks. Something in your brain will tell you it's going to rain exactly at the time you're planning a run, so let's postpone it for an hour (assuming people still need physical exercise in this form). As is already a reality today, networks of algorithms will steer the global market economy in ways that far surpass individual understanding.[2] As today, we won't notice much of it. These processes go on in the background of our perceptual awareness; they almost never come to the fore. Who knows to what extent our everyday lives will be affected and steered by these hidden infrastructures? The stocks of a certain automobile company have decreased and suddenly you find yourself with the urge to buy a new car. Some levels in your blood have gone awry, but before your body can develop any symptoms of disease, the nanobots in your bloodstream have already solved the issue. A network of sensors, distributed throughout the city, detects the gang of thugs (assuming crime is still a profitable business) waiting around the corner so it directs you unnoticed into another street. Your brain implant does that, remember. But you feel just a vague inclination. You cannot be sure if it's your decision at all—but whose else would it be? The technology's? *Where is* technology?

Technology is making itself invisible. It is evolving itself out of existence, that is, from a human-perceptual point of view—or what we still identify as one. Technology as we define it now, as we picture it, is developing in such a way that at some point in time, it will stop being what we still picture it to be; if it hasn't already. We will experience technology rather as *ourselves*, as *us*.

So here is a problem. We figure technology is *something*, something we can point to. But let's "intrapolate"—as the reverse of extrapolate—from these projected future developments: if technology tends to become something that we do not recognize anymore *as* technology, then shouldn't we start finding a new definition for it, now?

Philosophy of . . . What?

Of course, saying the word "technology" *is* already pointing at something, to be designated as technology. This may seem overly tautological. Nevertheless, it attests to an inherent problem in the scholarly domain known as the philosophy of technology. In fact, if anywhere people have tried to define technology in ways that surpass simple common-sense interpretations, it is in this field.

Basically, philosophy of technology in its contemporary guise attempts to find a middle road between the two "classic" views on technology: instrumentalism and determinism. The former sees technology as merely a means to an end: an instrument with which we aim to accomplish certain goals. And the goal-setting is up to us, humans. In itself, according to instrumentalism, technology is goal-neutral. This is illustrated in the famous motto of the National Rifle Association: "Guns don't kill people, people kill people."[3] Guns in themselves are value-free. Humans decide what they do with them, for better or worse. Not so for determinism: here, technology takes on an all-encompassing character, becomes something that pierces through to every realm of life and society—a force, power, or principle with its own agenda, that mostly does not augur well for humanity or for human qualities. For a determinist, guns harbor an intrinsic orientation toward violence, all good intentions notwithstanding. Phrased otherwise: a world *with* guns differs completely from one *without* them.

Contemporary philosophers of technologies find flaws in both arguments. Technologies are not goal-neutral: they push us toward specific uses, they "incline."[4] But they cannot be identified with plain, linear causes of societal effects either, like determinism suggests. Technology counts for one factor among many, and above all it can be modified, protested, controlled. It is "malleable."[5] In sum: technologies "do" things,[6] but their effects are not completely beyond our power. Interestingly, this makes both instrumentalism and determinism true up to a point. Determinism is right because technologies *have* effects. Instrumentalism is correct because human actors retain at least *some* control over what technology does. Extending the gun example: we may argue that since weapons are at least inclined toward violent practices, a society in which guns are not easily available will be a more peaceful one, but for some purposes (e.g., recreational activities, police work) the use of firearms may be warranted.

Historically, instrumentalism and determinism are deeply rooted in our cultural consciousness. Instrumentalism can be regarded as the oldest view, stretching back at least to nineteenth-century positivism. With the Industrial Revolution in full swing, and its side effects—socioeconomic misery, pollution, resource depletion—still dwindled by magnificent achievement and promise, the more optimistically minded could identify technology quite effortlessly as uncompromising boon. But then the twentieth century kicked in, with its horrendous world wars, the Holocaust, and nuclear bombs, and this substantially changed the atmosphere for thinking about technology. Thus, determinism came into its own by the middle of the twentieth century, with authors such as Martin Heidegger, Jacques Ellul, Karl Jaspers, and Ernst Jünger delivering scathing critiques of technology "as such." Technology became a threat, a force that escapes human mastery: something bigger than us.[7]

These classic viewpoints are still with us today. Vestiges of them can be found in our habits, practices, discourses, and cultural images. We often go about using technology as if it were merely a neutral means to an end. This way of looking aligns for a good part with that common view of technology as a "something," with which we set out to do "something." The newest digital gadget comes out and we rush off to the stores, expecting the device will help us to live more efficiently, comfortably, pleasurably—without a care for potential harmful consequences. Will it distort my social life? Make me sick with its radiation? Does its production involve miserable socioeconomic conditions for people in Asia? When I discard the device, what will happen to the waste? And so on; none of these questions usually pops into our head when we are standing at the checkout register. At the same time, visions of an all-conquering technological complex abound in our cultural imagination, most noticeably in cinema. Just watch movies of the dystopic genre, such as *Children of Men*, *The Road*, or *I Am Legend*; see how technology plays a kind of shadow protagonist there. Technology (or, more generally, technological-scientific progress) is often staged as main trigger for the installment of a gloomy post-apocalyptic era, that throws the characters back into a purportedly technology-less state. Once there, they are forced to fall back upon their supposedly natural wits, so as to overcome danger and build the human world anew. Here is visualized quite lucidly the determinist motive: technology is overwhelming "us," and we need to reclaim our pure humanity. What a difference with our behavior in the media store (or any other store for that matter), where we eagerly enough embrace our latest digital prosthesis: tablet, smartphone, smart TV, and so on.

This is such a schizophrenic situation. On one hand, we take technology for granted as a simple tool. On the other hand, we seem to want to process and parse its hidden conflictual potential—as in dreams—by telling and watching stories of technological calamity. Beneath the surface of our common-sense instrumentalist understanding an *itch* remains, unconsciously signaling to us the "rest value" of technology. But since we cannot digest this in our consciousness, as it doesn't tally with our assumptions, we need to revert to other, cultural means.

As said, contemporary philosophers of technology take the middle road. They attempt to draw a less diffuse picture of technology. It is the case that in everyday use, technologies often *appear* as just efficient means. This is how they, phenomenologically speaking, are disclosed to us in the first place. But that is not the end of it. We may peek beyond the mere appearance of efficiency, toward the wider effects and impacts that technology has. Quite a few representatives of contemporary philosophy of technology have in one way or another outlined such a dynamic: a dichotomy between a "narrow" view—technology is about technical efficiency and nothing more, or so it *seems*—and a "wider" perspective—technology fans out into countless complex networks of causes and consequences, and we can and should map those networks. I will call this in what follows and for the purposes of my argument the *central dichotomy* in the philosophy of technology.

The dichotomy already goes a long way in subverting "technology-as-something" thinking. Indeed, when you point beyond technology-as-just-efficient-thing toward a broader constellation of interacting elements, the technology-as-something (i.e., a thing to point to) partly disappears. In that capacity, philosophy of technology seems perfectly tuned to the technological developments toward invisibility. The domain in itself develops a *disappearing notion*. But at the same time, there is a big irony: in doing its job, so to speak, the philosophy *of technology* also partially annuls its very object.

Looking at Technology Anew

Of course, philosophy of technology's disappearing maneuver has a positive value. It signifies technology's "true nature": technology no longer is, and never actually was, a simple something. Any technology is spread across its networks of effects, other technologies, political-economic-social-cultural factors that constitute it as the technology that it is. We just get it wrong in everyday life, with our common-sense instrumentalist notion. We don't "get" technology. Sure enough, some insights of philosophy of technology have begun to seep into the collective consciousness. We are starting to learn, for instance, how a plastic container is never just a plastic container: it may harbor harmful substances such as bisphenol A. We are gradually mapping some technologies' larger-scale ecological effects. Yet the uptake of this kind of perspective remains slow and ineffective, and counterexamples are legion. We steer our cars roaring down the road, largely unmindful of environmental damage, the well-being of pedestrians and bicyclists, our own health. We use social media mostly unthinkingly, unaware of, or indifferent to, privacy concerns. We keep stuffing ourselves with industrially produced foodstuffs, possibly filled with pernicious ingredients. And so on.

Now one explanation for this latency in understanding is that we are still coming to grips, in our common consciousness, with the advances in intellectual-philosophical history achieved throughout the twentieth century. Like many other twentieth-century philosophical movements, philosophy of technology started out reacting to our modern philosophical-cultural heritage that circles around the Cartesian notions of subjects and objects, and a strict split between those two. Steeped in that worldview, we learn to see ourselves as subjects—autonomous, independent entities—positioned over and against objects that we perceive—regarded just as much as self-contained entities. Philosophy of technology teaches how technological things are never *just* things. Subjects and objects, with a term used in postphenomenology, "co-constitute" each other. But we still have to catch up with that new view. In everyday life, our modern legacy keeps working its influence, and we have a hard time shaking the epistemological habit of looking at technologies as nothing but freestanding objects that we, as sovereign subjects, effortlessly manipulate and command. The wider mode of grasping and understanding broader impacts, *beyond* the narrow, efficiency-oriented view, remains extremely inaccessible in practice.

However, if this would be the only issue, it would be solely a matter of integrating the insights from philosophy of technology more into our worldview, or waiting until this happens. Yet there is more in play. As stated, approaches in philosophy of technology elaborate a different view, in opposition to the modern legacy. According to these views, the human being and technology are no longer to be seen as two independent entities: they constitute each other. The human and technology are interwoven, ontologically, from the start, these views assert. Indeed, we can see a logical connection with their *wide* definition of technology: if technology is spread out to such an extent, in the end it becomes impossible on a fundamental level to distinguish between "us" and "it." Thus, we need new postmodern or amodern ontologies that radically go beyond the Cartesian subject-object split. Scholars such as Don Ihde, Bruno Latour, Peter-Paul Verbeek, and Bernard Stiegler have worked out such perspectives, based in different traditions.

Nevertheless, no matter how amodern these ontologies are, technological developments continue to challenge our idea of what it means to be human. This we notice on a day-to-day basis, when we (are forced to) face questions like: should we approve of GMOs? How far should reproductive technologies interfere in the "natural" reproduction process? Should we seamlessly mingle with brain implants? Et cetera. The amodern ontologies only go so far in helping us settle these disputes; we still need to *engage into the discussion*. Phrased differently: even though we are already *ontologically* merged with technology, we still need to make decisions about the extent to which, on an *ontic* plane, we wish to keep fusing with technologies.[8]

So here is a paradoxical situation. We ought to learn how technology is not a something but in fact is dispersed throughout a network structure, even to such a degree that we ourselves are part and parcel of that structure just as much. All in the same boat. Simultaneously, we're pushed every day to think about and decide upon specific instances of merging with technology. So we remain powerless with the common-sense instrumentalist notion that cannot account for the condition of *being merged*. But we stay just as powerless with the amodern "interwovenness" perspective of philosophy of technology, that does see humanity and technology as ontologically merged, but remains somehow blind to the process in between, that is, of merg*ing*. Indeed, we easily recognize this conundrum in the public debate about particular technologies or media, such as artificial intelligence (AI), machine learning, or self-driving cars. Typically some parties in the debate warn: "Watch out!" While others reassure: "No worries; technological evolution belongs to what it means to be human." Such discussions invariably arrive at a stalemate. Opponents see a certain technology as plainly incompatible with some essential human aspect. Proponents find no gap between the technology and the human being; however, we may have to mitigate nefarious side effects in an initial stage (that is the price we pay for inevitable development and progress). In the meantime, the technological developments just march on, while all the while everyone is suggesting: "We should investigate this, we should investigate this now, before it's too late!"

But this attests to the profound inadequacy of discussing these matters in terms of technology *x* or *y*, or in terms of something that we humans do. We need to start thinking differently about technology—not as a this or a that; but also not as an "us" (at least not in the way we are doing it currently). We need *another* lens. Now when a problem seems so convoluted, that must mean the solution is equally complex. Or that is what we would expect. But sometimes, solutions lie closer to us than we would think. In this case, we can find the solution (if such a term applies) surprisingly close to home, if we basically choose to revisit another common-sense idea about technology, namely, that we use it *for* something, for a certain *purpose*. Again, what could be more obvious? Yet philosophy of technology, as we will see, has in an essential manner forgotten about *purpose* along the way. To be more precise: it has relegated purpose to the narrow side of the central dichotomy, limited it to that side only. In everyday life we use technologies to achieve a purpose—better, faster, more cheaply, et cetera, thus, according to the dictates of efficiency. This lines up with the classic Weberian understanding of technology as purposive-rational action. But as we saw, there is *more* beyond this: wider consequences and effects. However, as soon as we enter that wider mode, we seem to have suddenly stopped talking about purposes—as if we can only make sense of "small," instrumental purposes, which are "on the other side." Yet there is a *wider sense of purpose* that we need to reinvigorate. But therefore we require a refurbished philosophical foundation or infrastructure. This, we will find, is actually already lying dormant within the philosophy of technology. But in order to fully bring it out, we need a conceptual key. That key, in this book, is the work of Gregory Bateson.

Gregory Bateson as Philosopher of Technology

He might seem an unlikely ally. Gregory Bateson (1904–1980) scarcely wrote about technology. He was also not a philosopher, but an anthropologist by training, and his subsequent career exhibits an eclecticism that dizzies even the most interdisciplinary mind. But he pretty much fell along the wayside in the pantheon of academic stars: his name doesn't really ring bells anymore (if it ever did) with a general audience.

Yet, although removed from the big spotlights, Bateson's place in intellectual history is indisputable. He made advances in domains as various as anthropology, cybernetics, communication, psychology, biology, and cognitive theory. Often the terms or notions he coined are better known than him: *schismogenesis*, information as "difference which makes a difference," *double bind*, "ecology of mind." He is seen to have been a trailblazer for ecological awareness. Since his death, attention to his work has been steady but not overwhelming. There have appeared a few intellectual biographies. The famous popularizer of physics Fritjof Capra regards him as an important influence, pointing out Bateson's relevance for a holistic, nonreductionist view of life. Others have stressed his importance for ecological education, while still others focus mainly on his theory of communication and its paradoxes. There has been continuing attention, up until this day, to his psychological work on therapy and interpersonal relations. Occasionally, someone adapts a Batesonian phrase or term, however, then the terms go on to largely shroud their origins throughout their further adoption, as in the case of Gilles Deleuze and Félix Guattari (*plateau*)[9] or Bernard Stiegler (*écologie de l'esprit*). Not in the least, through the writings and work of his daughters Mary Catherine and Nora Bateson, his legacy is kept alive.

But with this book I argue that beyond those readings, Bateson deserves an even more pronounced, more consistent stature as a great thinker for our time. In order to really unearth the grounds for that designation, nevertheless, we need to approach his work from another angle: it must be made clear how Bateson can help in confronting our day-to-day existential struggles. Those fan out into globe-spanning issues, and conversely, our global challenges trickle down to everyday life. At the junction of these concerns lies, indeed, *technology*—from our small daily dealings with smartphones, social media, financial transactions, transportation, health, and work to the big questions of automation, social injustice, climate change and power monopolies. All are mediated—and connected—by technology. Bateson's work, I believe, harbors unique perspectives to frame technology-related problems from a refreshing perspective. Nonetheless, bringing these out requires a special strategy. Judging superficially, one observes that technology plays just a marginal role in Bateson: he only sporadically mentions "technology" literally. Yet in fact the question concerning technology, to borrow Heidegger's phrase,[10] lurks in every nook and cranny of his thinking. Now that we have the insights from philosophy of technology at our disposal, we possess the right instrument to truly articulate this—by reading Bateson as a philosopher of technology. And at the same time, we can gain a fresh perspective on philosophy of technology by reading it through a Batesonian lens.

For Bateson, technology is first and foremost a matter of purpose. More specifically, it is the—often material—enhancement, or extension, of what he terms "conscious purpose": the typically human drive of wanting to accomplish aims. But conscious purpose in Bateson's account stands for only a small part of the whole "ecology of mind." The larger network of patterns that connect human beings to other living organisms far surpasses our narrow focus on goals. At this point a certain ethics or cultural critique steps in: we have gone too far in pursuing conscious purpose, Bateson argues, and have thereby started to neglect the true, that is, interconnected nature of the world. Yet that nature "works" in certain ways, and to understand how exactly becomes the main task Bateson takes upon himself throughout his career. The end result is an intriguing, though not completely finished—Bateson was in the midst of systematically synthesizing his views when he passed away—picture of a two-sided reality in which a "flux" of processes is punctuated by "difference."

I want to take these three aspects of Bateson's thinking together—technology analysis, ethics and cultural critique, ontology—in an endeavor to disclose Bateson's thought for the purpose of understanding our everyday involvement with technology. This is a particular effort not before undertaken in this form by any of his commentators. So at the same time I offer a presentation of, and introduction to, Bateson's work that can help contemporary readers get acquainted with his thought. That thought is systems-oriented through and through—Bateson being of course one of the pioneers of cybernetics. But he elaborated possibly one of the most holistic versions of systems thinking, tying together cultural, psychological, and social perspectives into one mind-dazzling frame. That peculiar combination of "big" and "small" can serve us well. Technology is literally ever expanding into wider concentric circles of world-covering networks, becoming ever more our environment: the water in which we swim. But it also makes the reverse movement, radically: into us, fusing seamlessly with our bodies, practices, habits. As we've seen, in that sense it is disappearing, if only from an everyday experiential-perceptual viewpoint. Bateson offers a systems-view perspective, however a peculiar one, that is perfectly suited to this situation: he eminently helps us to see systemic relations that might at first be invisible or hard to spot.

Across the Gap of the Central Dichotomy

Bateson's framing of technological issues in terms of purpose can, exactly by overturning the premises of the debate, cast our problems with regard to technology in a liberating light. Suddenly one is freed from the myopic focus on technology, able to look at issues from a broader angle. Specifically, this new perspective can also help us to take on some of the more fine-grained conceptual matters with which the philosophy of technology as a domain is currently struggling.

Philosophy of technology emerged as a full-blown philosophical sub-discipline in the last decades, and since the famous "empirical turn" of the 1980s and 1990s it has received a significant boost.[11] The field showcases diversity and richness, with various approaches each digging into their own conceptual roots, such as postphenomenology, critical theory of technology (also now called critical constructivism), and philosophy of information.[12] All have one methodological guideline in common: they take technology as prime point of entry for their study of the ethical, existential, social, or political characteristics of the *condition humaine*. As we've seen, this is already problematic in itself. But there are some more difficulties that the field is facing.

For one, the question recently arose whether philosophy of technology has not remained, or again become, *too instrumentalist*. Indeed, in distancing themselves from purportedly determinist approaches such as Heidegger's and Ellul's that see technology as an all-encompassing, all-penetrating force, contemporary philosophers of technology have made a heartfelt plea for the malleability of technology. But this aspect has perhaps been taken too far, with scholars expecting too much from technology's instrumental room for maneuver. Moreover, the times may be asking for a—at least partial—return of the notion of technology-as-overwhelming-force. At stake here most importantly is the increasing *algorithmization* of life and society. Through the growth and further mainstreaming of types of information and communication technology (ICT) such as apps, bots, wearables, sensors, and cloud services, more and more domains of life are being "algorithmized." That is, mediated and steered by algorithms: markets, administration, policy, health, education, social interaction, security, transportation, and so on. In the shape of the already relatively mundane phenomena of, for example, social media newsfeeds, on-demand streaming, and computer-generated news messages, algorithms are having a direct influence on everyday life. And with the growing importance of the "quantified self," "quantified other" and

Internet of Things (IoT), this impact should only intensify. Nevertheless, as already suggested, much of these networks of data, artifacts, and other components remains unnoticed, behind the scenes. The rapid growth of this hidden mediatedness of our world makes ICT's actual scope and impact increasingly escape our perceptual and intellectual grasp. An overly instrumentalist approach risks losing sight of this. These phenomena raise the long-standing question of human control over the technological entity anew. We have not quite yet arrived at *Terminator*-like scenarios, but algorithmic structures at times exhibit a behavior that can be, at minimum, perceived as autonomous and all-encompassing. Do we need to cast technology in a more determinist or essentialist light again? If so, how to do this without relapsing into pure, massive determinism?

In line with this concern about instrumentalism, there has been a debate going on about the status of the *empirical turn*. Some facets of technology, it is said—political conditions, for instance—can scarcely be scrutinized in an empirical manner. Empirical turn perspectives are currently being assessed for perhaps having become too empirical,[13] and calls are made for a return from the empirical: a re-transcendentalization or a "transcendental (re)turn."[14] Critics are concerned that too much focus on how technologies are experienced in specific, practical contexts—although looking at this is surely a commendable endeavor—may entail a dangerous disregard for their "system" characteristics. This may eventually cede the playing field completely to corporate and established political interests. The new developments in ICT described above make this lack even more pertinent. Also, there is a concern about methodology. As Dominic Smith argues, the heavy emphasis on empirical case studies enables a certain type of analysis but may block out others.[15] The very character of a case study makes some phenomena more suitable for study than others. Often, Smith suggests, the factors that determine and limit the choice for a particular phenomenon will not be made explicit.

These problems have fundamentally to do, of course, with that big ontological question about where the human being ends and where technology begins: is there a distinction between the two, or are they continuous? We saw how philosophers of technology in recent years have converged on the latter point of view, in opposition to modern ontologies that pit a subject (human, spirit, mind) against an object (nature, matter). Amodern ontologies see the human being and technology no longer as two independent entities, but as constituting each other. Examples include cyborg theory, actor-network theory, and postphenomenology. But as stated previously, how to rhyme this so-called human-technology continuity (or in Bruce Mazlish's words, "fourth" continuity)[16] with our daily struggles and controversies on "where to draw the line"?

It should be clear by now to the reader that these problems not only are interrelated but have essentially to do with the main issue that I described above: *to study technology is to study a vanishing thing*. In everyday coping or common use of technology, this issue manifests itself as a kind of conceptual reification: we understand technology as a thing, whereas it is—or at least is becoming—*more or less than* a thing. In philosophy of technology, subsequently, diverse approaches try to make sense of this situation, through investigating the network-like character of technology, but in the end they *lose*, exactly, their object of study. It will become evident in this context that philosophy of technology's fundamental problem in relation to the aforementioned issues is epistemological in nature. The field has been successful in pointing out side effects, ethical consequences, and to some extent the sociopolitical construction of technologies, but it does not yet pierce through to the grounds of the blindness for technological matters that still reigns in everyday life. Bateson's work, because of its emphasis on processes of knowing—human and other—can make a founding contribution here.

The crux of the matter, as we will see, is the central dichotomy. In its most fundamental form, the dichotomy distinguishes between what we can designate as a discursive and a material mode. As mentioned, in common use, we regard technology as "just efficient." This is the discursive side. Here we *talk* about technology strictly in those terms: it is an *x*, with which to achieve *y*. We could also call this the instrumentalist mode. Philosophy of technology, as we know, endeavors to go beyond that: it wants to map the constellation of side effects and impacts. Beyond our narrow discourse about technology—technology is just about efficiency—we must observe how in a broader perspective, it flows over into all sorts of *other things*, artifacts, procedures, processes, et cetera. On this side, the "material" side, technology does have effects. Hence, we could designate this perspective as the determinist mode. Searching for a middle road, philosophy of technology links up the two modes, tracing the contours of their connection, instead of merely attending to one side, like the classic views would. Simplistically speaking, then, the core argument goes as follows: technology is more than we say (or think) it is. Beyond our common discourse about technologies, in material reality they are *much more*.

In this context, some have spoken of a "material turn" (dovetailing with the empirical turn). Later on, I outline how, currently, authors are already extending or surpassing this material turn, seeking to bring to light what the turn forgets. Verbeek, for one, makes a plea for "one more turn" after the material turn, not only to understand technological, that is, material things, but also to make sense of how humans appropriate technology.[17] Mark Coeckelbergh wants to refocus attention on discourses about technology, as these essentially shape how we understand it, and he goes on to meticulously analyze how our ideas about technology are still shot through with romanticism.[18] The problem with these—in themselves highly worthwhile and relevant—efforts is that underneath them, the central dichotomy in its starkly dualistic guise is still lingering and doing its work. Put otherwise: the distinction between discourse and matter (or action) remains in place as a taken-for-granted building block. We will see how Bateson's unusual ontological-epistemological framework helps to unlock this tangle. For him, in fact, a similar dichotomy takes central stage: the distinction between mind (i.e., "discourse") and matter. But his whole work is bent on bridging the gap between the two, on describing what happens between the two sides. His notion of technology as a "materialization" of conscious purpose will be of cardinal importance here.

The Art of Living

My final goal will be to arrive at a framework that allows us to imagine technology—in ordinary use as well as in theory—in a truly different way. Bateson's thought can facilitate such a different way of looking. This exercise is not meant to stay confined to conceptual realms. At base level, Bateson's thinking was always oriented toward everyday practice—although he himself may not have made that sufficiently clear (this facet is actually more pronounced in the writings of his daughters Mary Catherine and Nora, who weave his work into their own), and in any case his practical orientation was at times ambivalent. He was wary, for a good part of his life—his attitude probably rooting in his World War II experiences—about any attempts to consciously change a skewed situation. That doesn't mean that his work didn't always ask the essential question: how do the "filters" before our eyes, the models through which we understand the world, shape our way of acting in it?

Bateson sees a way out, eventually. We can to a certain extent become conscious of those mechanisms and dynamics that mold our perception—the glasses that we wear on our nose, that distort our view. *Art* is the key word. This is meant in the literal sense of art as artistic practice, production, and consumption, but even more in the abstract sense of "playing with perception." Bateson develops the notion of art as a skillful navigating of epistemological waters that can offset the maladies of conscious purpose.

Out of my rereading of Bateson against the background of contemporary philosophy of technology, I want to distill a framework for an *art of living with technology*. All those three terms—art, living, technology—have a special poignancy in the Batesonian vocabulary. Bateson's primary object of study are living beings. He maps the ecology of mind. Mind makes living beings stand apart from nonliving things. Technology, then, is in his work the seldom-named but ever-present shorthand for the (cultural) phenomenon of conscious purpose being extended, enhanced, and sometimes driven too far. Art, finally, can offer the remedying perspective but never in a clear-cut, "efficient," solution-to-a-problem way. All three words in the phrase have equal conceptual weight. Nowadays, we more than ever need an *art* of *living* with *technology*. We must not focus on *technology* in itself too much, for the reasons described above: reifying technology makes us ignore its disappearing character. On a more mundane level, we might forget that our relationship with it takes the form of an intense, day-to-day involvement.[19] But just *living with* technology may not do either: we would run the risk of simply going with the flow, using technology like somnambulists turn on the lights.[20] So we badly need that *art* component.

What could an art of living achieve? Googling the phrase brings up the most various connotations, from yoga and meditation to "lifestyle" concerns (are the colors in my house balanced?) to cultural consumption. Here I am concerned with the notion that has been a longstanding staple in philosophy. In antiquity, the search for an *ars vivendi* was almost synonymous with philosophy per se. Stoics, Epicureans, Cynics, Skeptics: all endeavored to offer practical guidance on living well, that is, wisely, beautifully, ethically, joyfully, elegantly. Before them, the Greeks had put forth the notion of "care of the self" (*techne tou biou*), which was certainly not merely about the individual "me," but aimed at an ethical engagement with others and one's environment. Throughout the Middle Ages, the "art" element receded a bit into the background but was picked up again in full force by Michel de Montaigne, and later by a host of nineteenth- and twentieth-century thinkers such as Nietzsche, Heidegger, Sartre, and Foucault. More recently, art of living (*Lebenskunst* in German, *levenskunst* in Dutch) has become a branch of popularizing practical philosophy, its best-known international representative probably being Alain de Botton. "Popularizing" here should by no means be equated with "less valuable." In a sense, all *ars vivendi* philosophers have reached out to wider audiences to provide advice and help. It is almost unimaginable as such that one should reflect about what it means to live a good life without reaching out to one's contemporaries for conversation and consultation.

Only rarely within any of these traditions, however, is technology mentioned. Of course, one must remark that only in contemporary times have the effects of technology really started to be felt in everyday contexts. So earlier ambassadors of the art of living may be forgiven for this retrospectively attributed shortsightedness. Within the philosophy of technology, in turn, there has been ample discussion of "the good life" in relation to technology.[21] That concept does incorporate many of the issues at stake in the art of living traditions. Indeed, at best, analyses along this line focus on what technology is, how it finds its place in society, how people relate to it.[22] At worst, however, the notion of the good life—if only in its subsequent societal or political adaptation—winds up being equated with a sort of clinical balancing act, in which technology figures as the balancing chord: it is simply there, and one cannot do much more than develop the skills necessary for crossing. As such, living with technology becomes more of a technological practice than an art. Something still seems to be missing. We seem to be in need of something more. At this point, the art component comes in. And at this point, Bateson enters.

Our times need a Bateson: a framework that can expose and overturn reigning epistemologies that keep us trapped in vicious circles. We live in an age of exponential changes, and these have fundamentally to do with *us*, with how we think and consequently act. Climate change and global warming are developments we are scantly able to correct. Notwithstanding overwhelming scientific evidence, we have the hardest time taking effective global political measures, changing our behavior individually, and then sticking to it generally. We easily succeed in being "green" for a short while, since it is fashionable; but to enact long-term, durable change is a wholly different thing. Bateson was thinking about similar issues already in the 1960s and 1970s. He talks about the inevitability of "runaway" processes. In a cycle of positive feedback, a stimulus will elicit a stronger response, in turn provoking an even stronger reaction. By now, such insights are fairly standard fare (they weren't in Bateson's days, and certainly not at the dawn of cybernetics), but one aspect is still often overlooked: the fact that these processes are almost impossible to transform. Even if one of the parties attempts to side-steer the pattern, the amplifying effect may still arise in different forms. Say I threaten to have a conflict with my neighbor over a practical matter—the height of our shared fence, for instance. I might decide to give in and bow down to his wishes. On the face of it, this would defuse the mechanism of mutual positive reinforcement. But here is the catch: this defeat might begin to fester in my mind. I might start to nurse a grievance, and this may subsequently influence my behavior toward my neighbor. He may unconsciously pick up signals of my unconsciously cherished grudge and interpret them as hostile attitude—again fortifying his frustration and thus extending the conflict. We would still be trapped in the cycle. Society abounds with examples of similar dynamics, runaway processes that we are hardly able to sidestep. There are the obvious, old-time examples of violent political and religious conflicts. But in a more contemporary vein we may enumerate things such as subprime lending, ever-increasing study loans, and, indeed, algorithmization.

What we need is a new pair of glasses, a new lens, to look at these phenomena. Such a project can start with a thought exercise. Imagine that all your life, you have been walking around with glasses on your nose that distort the way you look at things. What are you imagining exactly? Are you watching yourself from a distance, seeing how you remove the pair of glasses, squint your eyes in dizziness, reach out your hands to find support as you struggle to hold your balance? Or do you imagine that you are moving through life with the distorting lenses in front of your eyes? Your imagined self in that case must surely be oblivious of the distortion. Remember, you've been wearing these glasses forever. How would you know they are giving you a skewed picture of reality? What you perceive is—as far as you are concerned—reality as it is. Let's go a bit further. Perhaps the pair of glasses distort *more* when you look in a specific direction. Let's say objects at eye level come in untarnished. But when you look upward or downward things get warped; of course you don't notice. So you've got a worldview that is accurate in some respects and erroneous in others—troublesome, because you don't know which parts you are getting right. I will not be telling you now you've been wearing spectacles all your life that mess up your clear perspective. "And here is the right way of looking at it." No, this book is about *imagining* it is so. What if . . . ? What if elements of our worldview are simply the wrong way of looking at it?

What if we are looking at technology the wrong way?

Overview of Chapters

In what follows, I want to imagine a new way of looking at technology—in that way opening up new, and hopefully finding more adequate, ways of existentially coping with it. Beyond the view of technology as a something, I want to explore another way of conceptualizing it. More precisely, I will conceive of technology *as* purpose. In "*something* thinking," we can only envisage a technology, *x*, meant to serve a purpose, *y*. We connect *this* technology to *that* goal or aim. What I am not after is outlining a program for finding new purposes for given technologies, which Andrew Feenberg aptly calls (in the context of a discussion on Marcuse) "a trivial idea."[23] No, I believe we need to reframe our definition of technology as such, start seeing technology as wrapped up in purposive structures, not only of the narrow, instrumental-purposive kind, but also of a wider systemic kind—the kind of purpose we usually overlook, forget, or cannot even fathom because of the concepts that shape our worldview. Beyond our "*x* for *y*" reasoning habits, we should be on the lookout for the value of *detours* and, as it will turn out, to that extent also the *beauty* of detours.

Such a project necessarily entails a moving beyond "something thinking," which is a move that philosophy of technology has already been elaborating and preparing,[24] however not sufficiently—given the persistent problems in the field, its inherent paradoxicality in regard to its object of study, and the issue of how to deal ontically with ontological human-technology continuity. Surprisingly enough, I will wind up eventually with an account of "objects" (namely, through implementing Graham Harman's object-oriented ontology) that nevertheless does not conceive of objects as we know them but lines up remarkably well with the relational perspective on *contexts* that Bateson delivers. The logical conclusion of my investigation will turn out to be that the central dichotomy of philosophy of technology needs to be leveled down to *all* contexts, *all* objects. It needs to be multiplied, reenvisioned as a multiplicity—just like technology is more and more "smeared out" these days, dispersed into definitely not a something but rather a "we don't know what."

Notwithstanding the tangibility of these issues—going through the world and looking at the world, we can easily recognize the dynamics at play, for instance of technology disappearing from view—this investigation is necessarily also characterized by a certain abstractness. It is about developing a new vocabulary, about training oneself to see through a new epistemological lens, the consequences of which cannot immediately be made concrete—as in learning a new language, one also has to find first a sufficiently firm footing in it, acquire some minimal level of fluency, before one can reap the fruits, really "feel" what kind of world that language brings forth. *The abstract* shapes and conditions how *the concrete* will appear. In fact, a reflection on this relation, that is, between the concrete and the abstract, is central to the project. My aim here first and foremost is to provide pointers, stirring readers into playing with their presuppositions. In the meantime, I offer a comprehensive introduction to both the contemporary philosophy of technology and Gregory Bateson's outlook. With all these ingredients in place, I am convinced a reflection is possible that is theoretical up to a certain point but that also has clear societal implications. If we really want to make sense of technology along the lines set out here, we can do this by way of *purpose*; by asking, actually, very simply, about things: "*What is the purpose?*" "*What's it for?*" We urgently, desperately need to relearn to ask that question, all the time—however, in a special, specific way, and that is what I set out to delineate. The notion of purpose itself will also need to undergo some multiplication. Purpose as we know it—as well as efficiency—will have to be manifolded. For beyond our "something thinking," there is always a "more."

I will proceed in three big phases. Part I lays the groundwork. In chapter 1, I sketch Gregory Bateson's thinking preliminarily in broad strokes, preparing the more detailed study further down the line. Chapter 2 puts in place the framework of philosophy of technology—which will serve as interpretive background for the rest of the study—offering a systematic description, as well as critical analysis, of the central dichotomy.

Throughout Part II, then, I lay out the Batesonian framework in detail, zooming in on the connection with technology. Chapter 3 first elaborates Bateson's epistemology, and by proxy of that inquiry, his ontology (the two cannot be distinguished for him). Starting from his theory about mind that puts all living things on the same level, he develops a picture of a two-sided reality. Bateson's work is on a fundamental level all about "twos." Having its own central dichotomy in this way, his approach can be seen to dovetail exquisitely with the central dichotomy of philosophy of technology. Next, in chapter 4, I consider Bateson's own views on technology, as far as he develops these in the context of his elaboration of the seminal concept of conscious purpose, and his critique of it. Technology according to Bateson amplifies conscious purpose—it is a kind of extension or enhancement of it. As such Bateson does not disapprove of technology; he is at the least ambivalent about it, but surely he sees it intensifying the nefarious effects of conscious purpose. The interesting thing is that Bateson can enhance our understanding of technology because his definition does not *start* with asking about technology, but with the question of purpose (whereas many approaches within philosophy of technology would rather go about in the reverse way, if they even bring in the notion of purpose at all). To further make this clear and prepare the actual mapping of the central dichotomy onto the Batesonian scheme, I delineate two other but related perspectives on purpose that will help us make sense of different kinds of purpose: Humberto Maturana and Francisco Varela's famous autopoiesis theory and Morris Berman's broad-ranging historical sketch of "participating" and "non-participating" epistemological paradigms. Subsequently, in chapter 5, the issue becomes: what to do then about the "problems" of conscious purpose? Here the paradox of conscious purpose surfaces: if consciously striving toward goals may have pernicious side effects, how could we consciously strive to do something about that without falling into the same trap? Bateson at first remains pessimistic about the possibility of there being any way out of this conundrum, but he gradually begins to sculpt tools that help to subvert the paradox. His theory of levels of learning (levels of abstraction) and especially his reflections on art—as a way of navigating ladders of abstraction and, crucially, as an activity or process contrasting radically with the *idée fixe* of conscious purpose—are of pertinence in this context.

Along the way, I already regularly refer to philosophy of technology, but in Part III the comparative synthesis of the Batesonian and philosophy of technology frameworks begins in full force—considering, of course, the issues and questions sketched above in this introductory chapter. Chapter 6 is concerned more with the nitty-gritty of the comparison between Bateson and particular strands and approaches within the philosophy of technology, but in the process it already starts to synthesize the acquired insights, providing a notion of *technology as purpose*, which unfolds across the central dichotomy in between narrow (that is, instrumental) and a wider (that is, systemic) purposiveness. To push this analysis further, and also to investigate how the relation between those two modes comes about, I eventually introduce an extra component in the shape of the object-oriented ontology of Graham Harman, whom I, unexpectedly perhaps, will treat as a philosopher of technology. In fact, Harman's perspective exhibits many similarities with Bateson's ontological-epistemological observations, but apart from that, Harman delivers us a key element—specifically in his treatment of the Heideggerian notion of *breakdown*—for the completion of the framework. In chapter 7, then, finally, I embark upon a deep reassessment of central elements of philosophy of technology, through the prism of that newly constructed framework. Throughout this, I draw the contours for an art of living with technology. More specifically, I thoroughly review the central dichotomy one more time, from different angles. First, I look again at the phenomenon of disappearing technologies—which can be literally experienced: technologies do become invisible, certainly in the context of algorithmization. I can now subject this disappearance to a more finely grained analysis, namely, along Batesonian levels of abstraction—which soon makes clear there is no "single" disappearance. Certain people or groups obviously have an economic or political interest in certain technologies or components thereof disappearing from view, experientially speaking; but exactly of this multiplicity of invisibility we can make sense by way of the Batesonian framework. This brings me, second, to a reevaluation of the relation between discourse and matter, a variant of the central dichotomy. Philosophy of technology has struggled with this relation, feeling forced to choose between the two. But actually the two belong together; just like the two modes of the central dichotomy should be framed as two sides of the same coin—and then this structure should be multiplied endlessly. There is no "one" dichotomy. This in turn, and inevitably, leads me, third, to a reconceptualization of the notion of efficiency, which, like the notion of purpose—and that is no coincidence—has been rather neglected in the contemporary philosophy of technology. Like purpose, "efficiency" can be a much more helpful concept if we can find an adequate way of relating to it, and that is, again, starting from the idea of multiplicity. Next, I return one more time to Bateson's cultural critique, to then finally wrap up the investigation by reflecting in a synthesizing manner on how we become *artful*.

PART I
LAYING THE GROUNDWORK

CHAPTER I

Gregory Bateson's "Ecology of Mind"

Away from the big spotlights of intellectual history, burrowed between tree roots, lies the work of Gregory Bateson—glimmering like a diamond unfound. Surely he is still well known. But he is famous in the way unfound diamonds tend to be famous: by name *or* by their ideas, without the two being connected in a systematic manner within the public consciousness. We may know the name "Bateson" and we may know some of Bateson's ideas, but we perhaps don't know the two are linked.

This might have a lot to do with the shape of Bateson's career. Trained originally as an anthropologist, he dipped his toes in several different disciplines. Remarkably, he helped to force a breakthrough in each. Cybernetics: Bateson was there at the Macy Conferences that shaped the field. Anthropology: together with his then-wife Margaret Mead he pushed the methodological and conceptual boundaries of that discipline. Psychology: he and his colleagues developed the notion of double bind and laid the groundwork for family therapy. He was a forerunner of ecological thought, one of those pioneers in the 1960s and 1970s who rang the alarm about things such as pollution, deforestation, and pesticides, when much of the world was still blissfully oblivious of them.

Nevertheless, this way of lining up his "successes" may be more symptomatic of our tendency, nowadays as strong as ever, to think in terms of neatly demarcated academic specialisms. In each of these, typically something like "excellence" is to be strived for. Such a view however doesn't necessarily serve to clarify Bateson's importance. He himself, although struggling throughout his career to find conceptual coherence, was largely immune to such policing of disciplinary boundaries. He was on the lookout for what he called, with Augustine, the "Eternal Verities": truths about the world and life that are valid across the disciplines.[1] Eventually, the methodological perspective (if this term is warranted) that for him came to aggregate most of his concerns was *epistemology*.[2] Under the banner of this notion, the seeming inconsistency of his work dissolves and morphs into something resembling a clear-cut story.

In this chapter, I introduce Bateson's outlook—sketching it in broad strokes here, to allow deepening along specific thematic axes further down the line. First, I briefly situate Bateson in his historical context; the word "context" as such proves to be crucial. Second, I list and discuss his most important concepts. Third, I investigate his legacy by way of a succinct survey of contemporary interpreters of his work, especially zooming in on media- and technology-related topics. Fourth and finally, I reflect upon acquiring the Batesonian perspective, a slightly experimental practice that we may exercise so as to really make the analysis that will follow fruitful.

Conscious Effort versus Context: Bateson's Career

Gregory Bateson grew up in an English family in which the scientific attitude was held in high regard and practiced fervently.[3] His father, William, was an established biologist who, by the time Gregory was born in 1904, had already been doing groundbreaking work on heredity and biological variation (most famously, he is said to have coined the term "genetics"). Being an atheist, William Bateson instilled his children with the spirit of independent research—probably not a trivial influence on Gregory, given the son's later wide-ranging palette of interests and unstoppable inquisitiveness. Gregory went on to study natural science and thereafter, not wholly to his father's liking, anthropology.

His subsequent working career reads like a postmodern novel. There just doesn't appear to be a clear narrative arc, a direction, or even a logic that ties together the different elements. The protagonist seems adrift on a sea of endless trial and error. At one time a spike in the narrative tension occurs, when *Steps to an Ecology of Mind* is published in 1972—a book that succeeds in eliciting some public attention.[4] But at heart, the volume simply reproduces the seemingly erratic sequence of his professional activities up to that point. To that extent it keeps confusing readers, even to this day. As said, it is as if Bateson dabbled along in several disciplines without much direction.

Strikingly, he appears to have looked at himself largely in that way, judging from Rollo May's comments: "He described himself as being like a cork carried along on the current of the stream."[5] Also according to May, he said of himself: "I have never made a free choice."[6]

Those lines seem frivolous, yet they are significant, attesting to a kind of worldview or perspective which for us westerners of the twenty-first century is hard to get our head around. They speak of a kind of resignation that is hard to digest. We expect straight story lines—we are trained in that way. Preferably, the hero in our stories overcomes hurdles and generally fights against the odds, by force of willpower, work, and perseverance. To hear someone claiming unreservedly that he has never made a free choice, even suggesting to derive some pride from that, is for us like hearing a joke (a split second later, one starts to hesitate: *is he serious?*). We want someone fighting the odds; not swimming or, better, drifting on waves of odds. We need our protagonist to develop plans and strategies for conquering and winning, not "letting it happen." Yet with Bateson, exactly the shape of his career—if this term is fitting—mirrors his thinking. Bateson is the personified illustration of *context*. (This in itself is an important Batesonian idea: a practical process should reflect in its form the conceptual infrastructure that it tries to bring into being. Thus, for instance, teaching about collaboration should be done in a collaborative way.)

We are so used to conceptualizing personal biographies in terms of conscious, planned, and intentional efforts that it takes a good amount of unlearning on our part to start seeing the interaction between those efforts and contexts—be they historical, social, economic, political, or personal. In fields such as science and technology studies (STS), scholars have since a couple of decades ago successfully put this unlearning principle into practice, showing how the "individual genius" image that is classically cast of scientists and inventors, is not the whole truth. Inventors' and scientists' achievements may just as much be the result of the right societal and social conditions as they may be of individual perseverance and brilliance. In Bateson's case, the biographical pattern seems so haphazard that it forces us to look at the contexts in which his work arose—for lack of a consciously strived-after program, or at least one that is immediately apparent; more on this shortly.

So, there is his upbringing in a science-oriented environment. He studied natural science and anthropology, then undertook several field trips to do anthropological research (some of these with his soon-to-be first wife, Margaret Mead). Patterns of interaction and communication already formed a central focus of Bateson's work at this stage, and they would return in his research on dolphins on the Virgin Islands (1963–64), for the Oceanic Institute in Hawaii (1965–72), his work with psychiatric patients and on family therapy in Palo Alto (1949–62), even in his wartime work for among others the US Office of Strategic Services (1943–45). But perhaps the strongest impact on his thinking was his participation in the round of conferences that began the cybernetics movement, from 1946 onward: the Macy Conferences. Cybernetics was from the start a multidisciplinary field—a perfect environment, so it would seem, for the intellectually voracious Bateson—aiming to understand how systems work. In the first instance the field investigated machinic systems, but several scholars quite quickly tried to transplant the analyses of classic cybernetic cases such as target-seeking missiles (cybernetics arose during World War II) to sociological, psychological, or even philosophical spheres. Bateson was one of those figures.[7] Cybernetic vocabulary and imagery provided him with just the right tools to systematically expound some ideas that were already brooding in his mind and that he had started to develop during his anthropological field trips.

For one, under the influence of the nascent field of cybernetics he was able to fully work out his account of feedback processes of social interaction. There is positive and negative feedback. In positive feedback processes, two parties reinforce one another's behavior. I get mad at you, and in reaction you become mad at me, spurring me to become madder at you, you then becoming even madder at me, and so on. Such a vicious circle or "runaway" process usually ends in bitter frustration, violence, or some hard-won form of reconciliation. In negative feedback processes, by contrast, the two parties develop opposite reactions. I get mad at you, but you find some clever way to deflect my grievance or to put it in another light (you don't "take the bait"), inciting me in turn to soften my disposition. The conflict doesn't get out of hand. Negative feedback patterns generally create more stable societal constellations.

In each new work environment, Bateson was picking up on new cues. And yet, notwithstanding the importance of the contexts in which he, like a cork on the ocean, finds himself arriving each time, there also begins to appear a *line*—the contours of a search, a quest. This quest takes shape through the perhaps unconscious act of stringing these contexts together, but the stringing soon starts to suggest a pattern. Indeed, there is a "drive" behind Bateson's thinking; I already mentioned how he was on a search for, as he eventually started to put it, "Eternal Verities." Once again, exactly the notion of *context* is central in this regard. Let's take a brief exploring look at his central framework.

Mind and Matter: Bateson's Framework

All across the apparent diversity of Bateson's professional career runs one thread: eventually he identifies the fundamental orientation of his work as *epistemology*. Bateson's epistemology is panoptic: it has an unusually broad scope, enveloping not just the human being, but all living things. All organisms according to him have mental capabilities: all have *mind*. Before one would suspect him of panpsychism, it is worth noting that Bateson's notion of mind is not spiritual but biological or "structural" in nature.[8] For a clearer grasp of what Bateson tries to point at, it helps to temporarily bracket one's common-sense understanding of mind as something exclusively belonging to humans and certain other mammals and related to the elusive concept of consciousness. In a sense, Bateson seeks to build the notion of mind anew, from the ground up. Mind in his view has more to do with the general act of coping in an environment. All organisms engage in such activity: the oak tree as well as the salamander, the amoeba as well as *Homo sapiens*.

What are the structural characteristics of mental activity that all living beings share? Mind, Bateson argues, basically entails the perception of differences: "Our sensory system—and surely the sensory systems of all other creatures (even plants?) and the mental systems behind the senses (i.e., those parts of the mental systems inside the creatures)—can only operate with *events*, which we can call *changes*."[9] Out of all events, changes, differences that an organism perceives, it needs to distill the important ones. These are what Bateson terms *differences that make a difference*.[10] Plainly put, they are the differences that urge an organism to make a change.[11] For instance, a gazelle may notice a lot of events in its environment that do not concern it in a direct way, such as the presence of other animals. However, if one of those animals happens to be a predator, for instance a cheetah, that specific appearance, or difference, for the gazelle becomes a difference that *makes a difference*: it urges the animal to flee.

Bateson equates the notion of *information* with exactly these differences that elicit an action or a change in whatever form on the part of the organism. Indeed, "mind" at the very least assumes the ability to handle "information" in this way. The definition, moreover, helps to point out that all living creatures do not just passively receive signals from the environment; they also respond to them actively. "The end organs are . . . in continual receipt of events that correspond to *outlines* in the visible world. We *draw* distinctions; that is, we pull them out. Those distinctions that remain undrawn are *not*."[12] Perception, although it depends on impulses coming in from the surroundings, is a creative process.

It would also be a mistake to regard Bateson on this basis as an idealist of sorts—as if "all" is mind. Matter still has a role to play in his perspective, but not as we know it. In essence, Bateson's project turns out to be—and he presents it thus himself—an attempt at solving the mind-body, or mind-matter, problem: to bridge the long-standing gap between idealism and materialism. The materialists especially are often a target of scorn in his work. Contemporary science in his view is riddled with an inappropriate version of materialism, given that it reduces all processes to matter—bundles or packages of atoms or other elements—and in tandem with that, to the equally undefined concept of "energy." He speaks of the "materialist superstition" that "*quantity . . . can determine pattern.*"[13] Phrased differently, materialists assert that quantity can *explain* pattern. They believe that processes—patterns, events—can be exhaustively explained in terms of material units, the connections between which can be expressed in quantitative form.

Yet the idealists are no better. They in turn nourish the "antimaterialist superstition" that "claims *the power of mind over matter.*"[14] To add insult to injury, contemporary popular instantiations of antimaterialism, specifically new age philosophy (Bateson is writing about this in the 1970s), unashamedly adopts terms from materialistic science, using notions such as "energy" in a "spiritual" interpretation—thus making the confusion complete. That these notions are mixed up should however not come as a surprise. The two beliefs are actually each other's complement. And this, Bateson argues, is also not a coincidence, since they originated in the same source, namely, the Cartesian dualism between subject and object, or between mind and matter. We can think of

> the mind/matter dualism as a device for removing one half of the problem for explanation [i.e., the mental, the subjective] from that other half which could more easily be explained [i.e., the material, the objective]. Once separated, mental phenomena could be ignored. This act of subtraction, of course, left the half that could be explained as excessively materialistic, while the other half became totally supernatural.[15]

Notice the imagery of *one half* and *another half*. This depiction of two halves is key to Bateson's work. Our intellectual history, especially since Descartes, has left us with a split-up world and worldview. Bateson's aim is to bridge the gap between the halves. He doesn't want to abolish the terms "mind" and "matter"; what we need is to acquire a new perspective on them. This is possible through the methodology of what he calls *double description*. The world—more precisely, the living world—is a matter of both spheres, mind and matter, interrelated in a specific way. To understand that specific way, to understand *relationship*, we must learn to see double—not reduce one sphere to the other (I can already remark how such a maneuver is highly reminiscent of Latour's mission to unmask modern "purification": the act of explaining phenomena in either social or natural terms—thus he also brings together two halves that were split up under the influence of Enlightenment philosophy).[16]

Bateson's oeuvre is filled to the brim with dichotomies and fields of tension: Creatura versus Pleroma, structure versus flux, quality versus quantity, map versus territory, digital versus analog, conscious versus unconscious, secondary process versus primary process, pattern versus process, calibration versus feedback. Some of these overlap, others partly overlap, and a few stand in a particular relation to each other. Throughout the chapters of Part II, most of these dichotomies will be discussed. But it helps to already keep in mind a general guideline: most of the time, Bateson is concerned with the *interaction* between the two poles of the opposition.

All of his diverse work in different disciplines can be seen to lead up to this point, of bridging the gap. In contrast to materialist and antimaterialist "one-half" views, Bateson sees *pattern* or *context* arising from the meeting between matter and mind, mind consisting in the act of making a "mark" within the senseless stream of material processes. But we shouldn't imagine this to be happening as something abstract, intangible, belonging to some remote metaphysical realm. This is the way living beings, you and me, *are* in the world. This is the way in which we communicate, communication centrally being about making marks.[17] And then more particularly, this is the way in which we communicate with our loved ones—the process going haywire if contradictory messages are given on a persistent basis, possibly leading to schizophrenic symptoms, as outlined in Bateson's psychiatric work of the 1950s and 1960s. A victim of such "double bind," often in family situations, is unable to distinguish between *contexts*, between levels of abstraction (much more on these in what follows); this brings about pathology.

Bateson's epistemological analysis does not stay confined to purely conceptual realms. What is more, it morphs, almost unnoticed, into a cultural critique. This aspect of his thinking is not always discussed, when scholarly attention remains restricted to his cognition theory (and his insights on communication, psychopathology, and therapy). Yet his critique of *conscious purpose*, as we will further investigate in more detail, retains much relevance for contemporary discussions on technology.

Conscious purpose concerns just as much the relation between two realms, between two halves: in this case—Bateson adapts the terminology from Freudian psychoanalysis—the conscious and unconscious part of mind.[18] Humans, of course, partake in both. Yet consciousness represents only a small portion of all our cognitive life: the tip of the iceberg. The unconscious component covers significantly more ground than the conscious, Bateson asserts (like Freud did). But for some reason, an imbalance between the two appears to have arisen. We humans, notwithstanding the preponderance of the unconscious part in our mental process, have come to attribute the greatest importance to *consciousness*—more precisely, to the conscious pursuit of *goals*. This is *conscious purpose*. The notion means to connote the focus on practical realization, that is, of plans, designs, and purposes that characterizes us industrious humans so well. Conscious purpose, one could say, is that which most of the time drives us (modern) humans forward in our everyday smaller and greater projects.

Yet it entails only a limited knowledge of the structure of the world that is in actuality much more comprehensive than our mere conscious-purposive perspectives can grasp. What is more, an all-too-stern attachment to conscious purpose makes us neglect the "cybernetic nature of . . . the world,"[19] a perception of which lingers somewhere in the unconscious. Such limitation holds a danger, Bateson claims: it may make us vulnerable to changing circumstances, since a monolithic focus on purpose severely strains our adaptive skills. There are ways out, we will see—art being an important one—but they are rough terrain. One precondition for these ways to "work" seems to be to assimilate to a certain extent the Batesonian perspective as such. I will immediately elaborate on this, but first I turn to some contemporary readings and interpretations of Bateson's thought, in order to assess his current legacy.

Relation and Communication: Bateson's Legacy

In her landmark book *How We Became Posthuman*, N. Katherine Hayles distinguishes between three historical waves of cybernetics.[20] Each wave is defined by a focus on a specific concept. The first wave, which started with the Macy Conferences and ran until about 1960, circled around the notion of homeostasis, the way in which systems keep themselves in a state of equilibrium. After that, more and more attention was given to the idea that the standpoint of the observer, who is watching the system under scrutiny, should be taken into account as well: reflexivity became the paramount notion in the second wave, which lasted until around 1980 (reaching its apex, Hayles suggests, with the development of the concept of autopoiesis by Humberto Maturana and Francisco Varela). Third-wave cybernetics, then, which is still the contemporary paradigm today (at least, Hayles writes in the 1990s), endeavors to synthesize strands of the first two waves, by outlining the concept of emergence and focusing on topics such as virtuality and artificial life. Hayles casts Bateson as a figure who was prominent in the first wave, but already foresaw the importance of the notion of reflexivity early on, and tried to convince his colleagues of it. Due to resistance from other important representatives (Warren McCulloch in the first place), the concept was not fully deployed at the time. Nevertheless, Bateson can be seen as a bridge figure between the two waves, although Hayles discusses him more as a protagonist of the first wave, not specifically as a representative of the second. She regards him as someone who speculatively worked out a framework of ideas that could at that moment not be tested empirically—which was one of the main reasons for skepticism from some of his Macy Conference colleagues, like McCulloch.[21]

Surely Bateson's speculations could count on more enthusiasm later on. Among his most avid champions has been the physicist Fritjof Capra, who in numerous publications, beginning with *The Tao of Physics*, stages Bateson as one of the most important sources for the new view of life and the universe that has been developed in recent decades in biology and physics.[22] Capra outlines a general shift of focus in those domains: from hierarchies to networks, from parts to the whole, from objects to relations, from things to backgrounds.[23] Processes, patterns, and environments take central stage; the "web of life," in his words, consists of a network of networks. Yet we humans have a tendency to structure that web as a hierarchical ladder; we are prone to reify, Capra observes. In nature there is no hierarchy, there are no "objects." Bateson in Capra's view is one of the best representatives of the relational worldview—someone who helps us understand the relationality of life and the world, but also our epistemological shortcomings in correctly evaluating this relationality: our inability to "see" it at first.

This relational perspective is probably also one of the central reasons for Bateson's enduring prominence in the field of media ecology. Media ecology is the body of research that grew out of the work of Marshall McLuhan and some of his followers. It sets out to investigate the interactions between media: their ecology. But that is not all: in the media-ecological account, media *are* ecological in themselves. With the terminology of the field, media are, and make, *environments*. No strict dividing line can be drawn between something that is a medium and something that is not. It is impossible to neatly circumscribe the "something" that a medium—or a technology, as McLuhan equates the terms—is supposed to be. Bateson's holistic notion of mind is often referenced here, together with his theory of communication and his definition of information as a difference that makes a difference.[24] Sometimes his work is discussed in tandem with Alfred Korzybski's general semantics, a sister domain of media ecology. This is no coincidental confluence. At a crucial moment in the development of Bateson's thinking—when he delivered the annual Alfred Korzybski Memorial Lecture of the Institute of General Semantics in 1970—he became influenced by Korzybski's central premise that "the map *is not* the territory."[25] This was instrumental in the further elaboration of his theory of learning, based in the analysis of levels of abstraction or logical types. A representation (map) of reality (territory) is always situated one level of abstraction higher than that reality. And all forms of communication are characterized by this dynamic. Bateson often uses play to illustrate: play is a message about an actual event, without it being the actual event. For media ecologists, especially the ones studying communication in particular, this constitutes an essential insight. Corey Anton for one has meticulously analyzed this aspect of Bateson's thought and others, for instance the digital-analog distinction.[26]

Some authors have discussed Bateson more specifically in the context of education and learning. Chet Bowers together with his colleagues updates Bateson's insights for the purpose of ecological education and ecoliteracy.[27] Henk Oosterling in the Netherlands refers to Bateson, and specifically his theory of schizophrenia and his ideas on paradox, among others when proposing and theoretically grounding his encompassing social project based in the city of Rotterdam called *Skillcity*.[28] Jan van Boeckel in his PhD dissertation on arts-based environmental education amply draws from Bateson's perspective on art and learning, putting it successfully to the test in his own artistic-educational practice.[29] In psychology, Bateson's legacy is of course still tangible, if only in the concept of double bind, and through his enduring influence on the field of family therapy. An interesting account of Bateson's theory of communication, fusing it with insights from

Taoism—a convergence already implicit in Bateson's thinking as such—and to which I will later return in more detail, comes from Douglas Flemons.[30]

Approaching most closely the focus of this book perhaps are a couple of readings that connect Bateson's work to topics of media or technology. Robin Mansell's *Imagining the Internet: Communication, Innovation, and Governance* deploys Bateson's views on paradox, levels of learning and adaptation in order to discuss and criticize the utopian ideal of a "cybernetic ecology—today subsumed within the information society."[31] Orit Halpern's *Beautiful Data* is an STS history of the use of data and ideas about vision and cognition in the twentieth century, which sketches the emergence of a new paradigm of rationality and cognition circling completely around the analysis of data, under the influence of, most importantly, cybernetics.[32] In this context, Halpern discusses Bateson, in particular his work in psychology, however more as an *exception* to that new dominant paradigm. Ian Angus, then, in his *Primal Scenes of Communication* puts Bateson to work for the proposal of his "critical theory of communication."[33] Angus describes media as "primal scenes" that not so much represent reality but constitute it as cultural-social-historical complexes. Within those complexes, Angus argues, consulting Bateson, we can find our bearings by becoming "reflexive" in relation to them.[34]

Bateson may not have the stature of typical academic giants, but clearly his legacy is still alive. What none of these publications seem to do specifically, nonetheless, is discuss technology in relation to his work.[35] This is what I will seek to do throughout this book, in studying the intersection of Bateson with the philosophy of technology.

Acquiring the Batesonian Perspective

There is something elusive about the Batesonian framework: something paradoxical in itself. You cannot "get" it until you grasp it; and you cannot grasp it until you "get" it. And even then, intermittently, you will probably lose your grip on it at moments when you cannot "feel" it enough. It seems as if you need to literally incorporate the perspective so as to pierce through to it. There is no purely intellectual, detached conceptual observation possible here. Such an observation would imply that you have your own conceptual models—looking at Bateson's from "your" standpoint, and turning over his ideas, as if they were a Rubik's cube. This won't do the trick. With Bateson, you have to *appropriate* that Rubik's cube first. He himself already foreshadows something of this sentiment when he observes in his "Last Lecture": "We face a paradox in that I cannot tell you how to educate the young, or yourselves, in terms of the epistemology which I have offered you except you first embrace that epistemology. . . . You must know the answer to your question before I can give it to you."[36] This is a kind of paradoxicality that Plato's Meno already faced when he asked Socrates how you can look for something if you don't *know* that something; and conversely, if you find that something, how will you know it is what you're looking for, since you won't recognize it as such?[37] Wittgenstein famously opens the Preface of his *Tractatus Logico-Philosophicus* with the seemingly mysterious notice: "This book will perhaps only be understood by those who have themselves already thought the thoughts which are expressed in it—or similar thoughts."[38] And indeed, more than half a century later Bateson writes in the same vein: "This book can tell you nothing unless you know nine-tenths of it already."[39]

It may appear a threatening admonition that makes the neurotically inclined freeze in their tracks. As if a special kind of knowledge is required first, in order to then, and only then, make sense of Bateson's ponderings. In fact, the opposite is the case. "Those who lack all idea that it is possible to be wrong can learn nothing except know-how," Bateson reassures us.[40] There is nothing esoteric about Bateson's perspective; it is educational to the core, open to elaboration and experimentation.

Also important to remark at this point is that Bateson's theory itself was never finished. He passed away in 1980, leaving at that moment a body of thought that he had only since a couple of years before been trying to mold into systematic form. The book *Mind and Nature* (1979) was his first published attempt at synthesis. A few years after his death, his daughter Mary Catherine completed the unfinished manuscript *Angels Fear* (which had been intended as a collaboration between father and daughter from the start).[41] In a sense, that book represents the ultimate, more or less coherent product of his thinking, albeit that it clearly still leaves a lot of dots unconnected. What we have in essence is a collection of conceptual dichotomies—the set of distinctions to which I briefly referred above; they do not precisely overlap, but at the same time they are also not mutually exclusive. We must work with these relatively loose threads. But even constructing an idiosyncratic tapestry making use of them could already pay off in conceptual-philosophical currency.[42]

As said, I will read Bateson in the context of the philosophy of technology. The key to opening up Bateson's framework to the philosophical study of technology might just be its paradox-like, make-your-hands-dirty, grasp-it-to-grasp-it, elusive nature. Against all-too easy, all-too-natural reification—technology is *this*, technology is *that*—Bateson requires us to rethink our epistemological suppositions, slowly acquaint ourselves with a radically alternative perspective. In times in which we risk to not know anymore *what* technology is or *where* it is, we might just be in need of a view that *does not know*, and refuses to know, in itself. And we will find that what we have here, essentially, is a view in which we seek for pointers and clues *in every new context*, over and over again.

CHAPTER 2

Philosophy of Technology

Philosophers are drawn to dichotomies, like moths to light. They adore distinctions—between two categories, concepts, modes. Imagine the history of Western philosophy visualized as a timeline chart of the type that one can spot in high school classrooms; how many dichotomies would not decorate it? Being-becoming, transcendent-immanent, reason-perception, essence-existence, and so on. Historical periods in Western philosophy can be fairly neatly distinguished by just looking at which dichotomies take central stage at a specific moment. "Fairly," I say—it is not exact science. Also, we famously overestimate the extent to which historical eras are consistent lumps that project a clear identity. Nevertheless, as Foucault abundantly demonstrates, there do exist *epistemes* that define a given age or field.

Along these lines, I argue, there is an *episteme* of philosophy of technology, and it hinges on one *central dichotomy*. Multiple scholars have sought to operationalize various dichotomies for the philosophical investigation of technology use. Within sets of these, there is more coherence to be found—like with historical eras—than has been made explicit up until now. Such endeavor I undertake in this chapter.

The central dichotomy, we will find, comes down to some extent to one more variety of another philosophical oldie: the word-thing dichotomy. In another form, it concerns a difference between *what is said* and *what is done*: the story told about what happens in the world on the one hand, and that which actually happens in the world (the "things") on the other hand. We don't need to get bogged down into long-standing and often not very constructive debates—onerous discussions about "truth," that purported holy link between words and things. Yet evidently we are dealing with epistemology. Moreover, *said* and *done* are not to be taken too literally. And we are not concerned with moral problems, at least not in a direct sense (e.g., do what you say you will do), although the cocky expression "You talk the talk, but do you walk the walk?" captures something of the broad conceptual sentiment we must eventually try to convey—even in the ethical sense.

First, I will dig into the historical and conceptual roots of the central dichotomy; that is, Heidegger's tool analysis. Second, I will outline and discuss a selection of instances of the central dichotomy in contemporary philosophy of technology. More precisely, I zoom in on three main approaches: postphenomenology, actor-network theory, and critical theory of technology. Third, I discuss the general characteristics of the central dichotomy. It will appear, then, that it harbors its own complications, which I elaborate in a fourth step. These analyses will set the stage for the remainder of the book, so in a fifth and final step I synthesize the findings of this chapter in order to lay the groundwork for the investigations to come.

Roots of the Central Dichotomy

One of philosophy of technology's main occupations has been to qualify the common-sense understanding of what technology is and does. That common definition of technology is—still today—instrumentalist in spirit. According to this view, technology is a means to an end, something we use to accomplish goals. Those might be new goals, goals that were hardly even imaginable before the advent of larger-scale technological complexes, such as satellite broadcasts or flying to the moon. Or they might be old goals that we now, with the help of technical devices, practices, or processes, are able to achieve easier, faster, cheaper, or on a bigger scale. Examples are (industrial) food production and sewage disposal. In any case, the instrumentalist perspective perceives technology as something that improves, betters our predicament; something that strengthens, enhances, extends our naturally given human capacities; something that alleviates or even eradicates flaws and hazards . . . and no further questions asked.

Martin Heidegger was one of the first to explicitly denounce such an interpretation.[1] Technology, he argued in his later work, more than being merely a means to an end, is the latest instantiation of Western metaphysics. Referred to by Heidegger as the "Enframing" (*Gestell*), modern technology comprises nothing less than the way in which being reveals itself to us in our time. In this capacity, it harbors great danger, standing for a form of calculative thinking that seeps through at every level of modern existence and that aims at the uniform summoning ("ordering") of all raw materials, including humans. Technology, in this view, is anything but a neutral tool. In fact, one could not be more diametrically opposed to the instrumentalist standpoint than Heidegger is in his study of "the question concerning technology."

So how did common sense wind up in the same corner with metaphysics? Nothing could be further removed from common sense than metaphysics, one would expect. Yet at this point we admire phenomenology's masterstroke: it sees everyday practice and everyday thinking as molded and maneuvered by metaphysics—however, without the former being aware of the presence of the latter in its very architecture. Everyday perception is rife with biases, rooted in a metaphysical worldview that has been handed over across the generations. We have learned to look at things in a certain way—the last big reference point for that historically conditioned perspective being the Cartesian subject-object division. We have grown accustomed, at least in the West, to look around in the world and view things as free-standing objects, just as we ourselves are free-standing subjects. But really this is just a mirage, phenomenology suggests: there is no neutral thing. Phenomenological method is exactly devised to bypass those filters before our eyes (and minds) and pierce through to untainted perception, to—in Husserl's iconic phrase—the "things themselves."

Heidegger doesn't stop there, however. He is adamant that in the meantime—we can "bracket" all we want—the history of metaphysics just keeps marching on. And it's getting much worse, actually. It is getting so bad that "only a god can save us."[2] For it is in technology that "subject-object thinking" is reaching its grandest, all-encompassing, and quite threatening realization.

Contemporary philosophers of technology still regard Heidegger's analysis of technology as seminal, if only from a historical angle. Yet most of them are significantly more interested in his earlier work, particularly the tool analysis from *Being and Time*.[3] Heidegger there introduces the illustrious distinction between two types of being, two states or modes that alternately define our outlook on the world.[4] On the one hand, there is *Vorhandenheit* or presence-at-hand, the objectifying, reifying attitude toward beings that has dominated metaphysics and is epitomized in the doings of modern science. On the other hand, there is *Zuhandenheit* or readiness-to-hand, the "actual" situation in which the objects surrounding us are not so much "placed before us" but are embedded in a "relational totality" ("equipment") that is defined and interpreted in terms of the work to be done (taking shape as a "toward-which" and an "in-order-to"). Heidegger's famous illustration concerns the hammer. When *zuhanden*, the hammer is not considered as an object, but recedes into the totality of the activity of for instance hammering a nail. The hammer disappears from "view." It starts to become *vorhanden* when, for example, it gets broken and we suddenly notice it, in all its deficiency, as a "hammer object" (more on this phenomenon of "breakdown" shortly).

Scholars nowadays tend to cast the tool analysis as more nuanced and delicate than Heidegger's later, rather monolithic, and seemingly dystopian technology critique. Yet in an essential way, the two relate. Western philosophy—our dominant mindset—frames beings as objects. In that vein, it cannot but see technology as a free-standing "neutral" entity. However, reality is really *relational* in character. Thus, we should adjust our definition of technology accordingly: it is more like a *system* or network than a tool or *thing*; more like a constellation of effects than just a neutral instrument. In this bare-bones version, extracted from its gloomy-dystopian frame, the Heideggerian *Vorhandenheit-Zuhandenheit* distinction survives up until this day and has been quite influential in the philosophical inquiry into technology. In one way or another, the most important approaches in contemporary philosophy of technology incorporate or extend it, either explicitly or implicitly. I discuss this in more detail in the following section, listing three of the main strands in philosophy of technology today: postphenomenology, actor-network theory, and critical theory of technology.[5] Why these three exactly? There are of course many more approaches available in the field. But it will become clear in a later stage that these three interrelate in a logical, specific way, and each have their particular emphasis, so that taking them together delivers us a representative "map" of the central dichotomy.

Instances of the Central Dichotomy

Postphenomenology

Of all dominant perspectives in contemporary philosophy of technology, the subdiscipline of postphenomenology has probably built most explicitly on Heidegger's tool analysis. It is also postphenomenology that, as the name betrays, builds on the tradition of phenomenology as such in the most direct way.[6] But the tool analysis occupies a special place in postphenomenology's conceptual toolkit.

Founder of the domain Don Ihde explains how Heidegger's analysis of "equipment" entailed a critique and inversion of Husserl's phenomenology.

While for Husserl piercing through to the essence of phenomena—phenomenology's core business, so to speak—happens through transcendental subjectivity, Heidegger sees something else as primordial. In dealing with the world, as said, we are always already engaged in a structure of relations, most often taking the form of practical involvements. This is the ready-to-hand. We are not first and foremost envisioning the hammer as hammer object. That latter, abstract mode of grasping a thing theoretically or predicatively is the present-at-hand.

The two modes, though, are not absolutely equal, Ihde helps to point out.[7] Equipment *precedes*. The present-at-hand derives from the ready-to-hand. And there are a couple of typical situations in which the ready-to-hand is forced, at least to some extent, into the direction of the present-at-hand—in which the hammer as object becomes visible. These situations are known as *breakdown*, with Heidegger describing three types of it: conspicuousness, obtrusiveness, and obstinacy. The hammer might be broken (conspicuousness), it might be lost (obtrusiveness), or something might stand in the way of using it (obstinacy). In each case, the fluent relational totality of praxical concern collapses and we are forced to attending to the hammer as "hammer." But ontologically speaking, this is not the primordial state: *first* there is equipment, *then* the present-at-hand arises from breakdown. We will see how this is largely convergent with Bateson's inherently relationalist standpoint: in his view, too, *relation* comes first (weirdly, however, we will also find Bateson to a certain extent aligning with Graham Harman's upside-down interpretation of Heidegger's tool analysis, that sees readiness-to-hand exactly *not* as relation but as substance).

For postphenomenology, the primordiality of relation is crucial. Zooming in on technology, postphenomenology deploys the methods of phenomenology specifically to understand how technology comes about in practical use. One outcome of this endeavor is Ihde's well-known "phenomenology of technics" that systematizes and analyzes different sorts of "human-technology-world relations."[8] It sets out to investigate which parts of a technology we experience as transparent and which as opaque, and how these ratios shift. In essence, Ihde's relations analysis can be seen as a concrete application of the tool analysis. In every type of relation, a certain configuration of *Zuhandenheit* in-relation-to *Vorhandenheit* is at work. So, in fact, what the "phenomenology of technics" shows is how the central dichotomy is at play across different ratios in different use contexts.

In *embodiment relations*, the technology or tool tends to be wholly transparent, while world (in the phenomenological sense of the term), on which I am focused, stays opaque. That means I see the things, I don't see through them. The clearest illustration are glasses: I wear the glasses, that is, embody them, incorporate them into my body scheme. And so mostly I do not notice them. I see through them and attend to other things. My intentionality is directed at world, while the technology and I become fused to an extent. Nonetheless, the technology is still mediating my experience—if only by enabling sharper eyesight.

In *hermeneutic relations*, subsequently, the visible-invisible border shifts toward technology and world. The technology here is used for me to interpret or "read" world. An example concerns the dashboard in the control room of a nuclear power plant. By definition I simply cannot perceive what happens in or around the reactor core, given the hazardous levels of radiation. So I read the situation on the meters and dials in the control room. Intentionality is now aimed at the technology (the metering instruments) through which I learn something about world (the core). But in a phenomenological sense, technology and world are scarcely distinguishable; they become merged. Here too, at the same time, the technology enables something that wouldn't have been possible without it.

With *alterity relations*, then, the transparency-opacity ratio is different again. Here I interact with a technology as if it were an "other," such as for instance with an ATM machine. World disappears almost completely from view, and my interaction is exclusively with the machine as such. The technology becomes opaque, the world transparent. In the case of *background relations*, finally, the technology becomes transparent, while world takes on an opacity again. The prime example is a central heating system, that just sits there without me noticing it except for maybe a hum now and then, or when the temperature is off. I just mind my business while these background technologies (another example is a refrigerator) do their work unnoticed.

These illustrations may be fairly simple and harmless, but the way of reasoning has been successfully applied by Ihde and several other authors to more challenging kinds of technologically mediated phenomena such as visualization technologies, phoning while driving a car, public benches as sleeping places for the homeless, digital media, and many more.[9] In each instance, the relation of visibility (opacity) to invisibility (transparency) and how this plays out in the particular context of use are under scrutiny. Essentially, we could repeat, postphenomenology takes Heidegger's across-the-board analysis of the two fundamental modes of being—the tool analysis—and looks for tangible versions of it in concrete circumstances of tool (i.e., technology) use.

This is also confirmed, albeit in a more negative manner, in postphenomenology's critical or at least ambivalent attitude toward Heidegger's later philosophy of technology. Ihde mentions approvingly how "[t]he analysis of equipment, the ready-to-hand/present-at-hand distinction, and the understanding of a praxical 'knowledge' in which tools 'withdraw' and yet remain 'assigned' to complex contexts, remains in my mind as one of the pioneer analyses of technologies in use."[10] Yet Heidegger's analysis of technology in "The Question Concerning Technology" can count on much less admiration. The latter's gloomy depiction of technology as a massive force, as the latest instantiation of metaphysics, that can only be countered by some kind of withdrawal into contemplative thinking, into simply *waiting*, represents for the postphenomenologist the perfect recipe for overlooking technology in all the places where it can actually be found: in countless practical contexts, that show technologies to be in fact highly malleable and open to interpretation. But even more fundamentally, Ihde argues, Heidegger's later technology analysis undoes the valuable work of the tool analysis: "In some sense, the illuminating distinctions of the ready-to-hand and the present-at-hand of *Being and Time* collapse in the later work and become unified."[11] Indeed, in an age in which technology becomes nothing less than "our reality," the calculative thinking that reduces every thing to an object placed before us in present-at-handness would become a "systemic" given. Because that kind of thinking would become all-encompassing, it couldn't be anything else than a relational totality of sorts. The two modes would fold into each other, and we enter a determinist perspective, in which technology is *all*. The original tool analysis, at least in the way postphenomenology interprets it, exactly allows us to see beyond what we think technology is and find out that it is *not all*.[12] This is made even more clear by the perspectives I discuss next.

ACTOR-NETWORK THEORY

For, remarkably, in its most generalized form, the central dichotomy can also be found in several other approaches to technology that do not directly regard Heidegger as an influence. Bruno Latour's actor-network theory (ANT) is one of them.[13] Although Latour may not be very fond of Heidegger, he does share to a large extent the latter's project of debunking modern metaphysics. Of course, Latour's take is different. In his view, as is well known, "we have never been modern."[14] Modernity is only the story we tell ourselves, the discursive veil—meant to make the room look tidy—that we throw over the rubbish scattered across the floor. All the while we keep believing we are experts in cleaning up. For, as moderns, we have gotten into the habit of categorizing all things in the world as belonging to either of two sides: society and nature, or in different terms, subject and object. We succeed in "purifying" the world in this way. Yet the rubbish doesn't disappear. In reality, we have never stopped "proliferating." That is: we have kept messing things up, crisscrossing easily between these so-called subject and object poles, making hybrids of humans and nonhumans, "quasi-objects."

This "Middle Kingdom" of actants enmeshed in networks is what Latour wants us to aim our attention at. This is where everything happens. We should be looking here instead of to the simplified version of events that the "modern Constitution" presents us with. That also holds for technology. The Moderns would claim technology to be situated clearly on the object side, something neatly separated from us subjects, and to that extent perfectly controllable and manageable. Evidently, this perspective must lead to an instrumentalist view. But of course there is more to it. Latour's revolutionary analysis of *closed* versus *open black box* dynamics makes it tangible.[15] In regular use, Latour avers, a technology acts as a black box that we do not question, except from a technical perspective. We just ride our bicycle without wondering how it ever came about, how it acquired precisely this form or this design, or what other designs could have been possible. We just use the tool, evaluate it on the basis of efficiency, of its technical characteristics alone: does it work like it should?

However, there are always other values than just technical values incorporated into a technology's design: social, political, economic values. And during a certain period in the past, these were still in flux, that is, under discussion, modifiable. In the case of the bicycle, this is famously illustrated by the research of Trevor Pinch and Wiebe Bijker, who have demonstrated how different designs for the bicycle were in use at the same time around the end of the nineteenth century.[16] The eventual choice between the "penny-farthing"—the bicycle model with the gigantic front wheel that we now know only from old postcards—and the so-called "safety bicycle," the model that "survived" and with which we are familiar today, was not a solely technical matter. It involved considerations of safety, culture, gender relations, economic factors, and social identity. But the success of the safety bicycle, and the apparent obviousness it acquired throughout the subsequent decades, has made us look upon its once-equivalent alternative as primitive; while the safety bicycle now has the air of being the most technologically advanced. From the moment a technology's design gets fixed, the other, non-technical values become cloaked by a mere focus on efficiency and functionality, and we forget its social genealogy. The black box is then "closed." Luckily, in studying the construction history of a technology we can "open" it again, lay bare its network structure, and see how it is incorporated in webs of values and interests.

The analysis of the tension between closed and open black box can be read as a more "practical" description of the more fundamental purification-proliferation dichotomy. Closing the black box stands for purification; (re)opening it is to recognize and account for proliferation. Interestingly, what counts for technology and science also applies to other domains of human practice. Graham Harman comments in his (first) book on Latour, *Prince of Networks*: "In a sense, all human activity aims to create black boxes."[17] Latour himself offers the following definition: "When many elements are made to act as one, this is what I will now call a black box."[18] So what appears as "one" stands on one side, but really that is just a reflection of the "many elements" standing on the other. Beyond a simplistic story, we can and must look for a more complex, network-like reality or history.

CRITICAL THEORY OF TECHNOLOGY

Another philosopher of technology, Andrew Feenberg, then elaborates upon this notion of the black box, and to a good extent also upon Heidegger's tool analysis, when working out his critical theory of technology or, with another denomination, his "instrumentalization theory."[19] He nevertheless goes further than Latour as well as Heidegger in also thoroughly analyzing the consequences in terms of politics and power relations.

"Instrumentalization" is meant to structurally describe the process by way of which technologies find their place in society. That process is inherently double-sided, consisting of two opposite movements: primary and secondary instrumentalization. The former entails the development and introduction of technical tools from a purely functional perspective, according to the principles of technical or instrumental rationality (Max Weber's *Zweckrationalität*). In that process, Feenberg argues from a Frankfurt School tradition, first and foremost political and corporate interests are served, in that technologies get to be deployed—often unnoticed—as means for the preservation and consolidation of the establishment's power. By contrast, secondary instrumentalization stands for the counteracting measures taken up by individuals or interest groups, which Feenberg also calls "democratic" or "subversive rationalization," the disruption of technical rationality and its attendant power-related biases.[20] Here technologies are incorporated in networks and applications, appropriated by ordinary users. It comes down to situating a strictly efficiency-oriented predisposition within the wider context of a social, lived environment.

In thus elaborating combined insights from critical theory and social construction of technology (SCOT), Feenberg argues that technologies incorporate nontechnical, that is, social values, but for the most part these get embedded in a form that *appears* as purely technical. That form he calls the technology's "technical code."[21] Exactly this hiddenness, he avers, enables technologies to be deployed by dominant groups to consolidate their power. Feenberg speaks of a "technological unconscious."[22] In the early stages of a technology's development, values are still discursively articulated—and discussed and fought over—but gradually they get "repressed" toward the technological unconscious, and henceforth appear as merely efficient, exclusively technical standards. "Often current technical methods or standards were once discursively formulated as values and at some time in the past translated into the technical codes we take for granted today."[23] Or, phrased the other way around, "values are the facts of the future."[24] Some of them are bound to be one day "cast in iron."[25] At that moment the "illusion of technical necessity" gets created, and "the waves close over forgotten struggles."[26]

Nonetheless, some interchange does remain possible: "Design embodies only a subset of the values circulating in society at any given time. Those not so embodied appear discursively rather than technically, but the two forms of value are not irrevocably cut off from each other."[27] On the contrary, in order to contest or modify a technology, the social values harbored by that technology need to be made explicit in one way or another. It is to be stressed that Feenberg argues that the "technological unconscious hides *the interaction between* reason and experience"—"reason" and "experience" corresponding largely with the aforementioned processes of primary and secondary instrumentalization, respectively.[28] Thus, there is still interaction. But the crux of the matter is that modern societies, unlike traditional societies, *hide* this interaction: "Traditional societies do not hide the substantive consequences of the exercise of authority. . . . But modern formal rationality serves similar social purposes under an appearance of neutrality."[29] Once one realizes how this dynamic works, however, one can begin to uncover the interrelation: "Once one grasps the force of the technological consensus underlying our way of life today, it is more plausible to imagine changes at that level winning widespread agreement."[30] But for that, social values need to be brought out in the open again. That is what Feenberg envisions with his program of "technical micropolitics," which can involve technical controversies, innovative dialogue, or creative appropriation.[31] In all cases, these initiatives are concerned with "cracking" the technical code.

Feenberg with his approach lands somewhere in the middle between Heidegger's purportedly essentialist view and Habermas's definition of technology as merely an "objectivating attitude,"[32] or, in other words, between the determinist and instrumentalist standpoint. Neither Habermas nor Heidegger really look at technology *from within*, Feenberg observes, but both views can be revitalized and adapted to our situation. Both are right to an extent; technologies appear visibly as neutral because of the workings of technological hegemony and technical codes, but they actually instantiate and confirm invisible power structures. Once again, we find a dynamic between the presupposition that technology is this kind of *all*; whereas digging deeper, beyond that, we can discover that it is far from an *all*. There is a world of *non-all* out there.

What Is Said and What Is Done

Generally put, in all the cases above, two mutually opposite views or attitudes are outlined. One view keeps to a purportedly limited perspective and stands for a certain narrowing or dumbing down of an actually much broader or richer situation (presence-at-hand, closed black box, primary instrumentalization). The second view, conversely, represents the "fuller picture" and is able to account for the multifarious and intricate reciprocal relations between human and lifeworld (readiness-to-hand, open black box, secondary instrumentalization). What is more, all these approaches implicitly suggest everyday practice to be tainted by a predominance of the first, narrower mode: technology use is aimed first and foremost—at least from an experiential point of view—at the efficient obtainment of specific results. It takes place under the auspices of the dictate of ("Weberian") technical rationality,[33] whereas actually it is at any time embedded in and determined by wider, elusive constellations of social, cultural, political, economic, and many other factors.

Surprisingly, the instrumentalist definition gets recuperated to a certain extent in each instance. True enough, from a phenomenological, practical, or user perspective, technologies are *experienced* as functional means, or at least we *talk* about them in that way. Yet Heidegger *also* had a point in radically going against the grain of that interpretation and posing technology, conversely, as a systemic, all-encompassing realm or "revealing" of being. The instrumentalist definition indeed does neglect the relational sphere surrounding, in fact constituting, technology. Heidegger, however, might have compensated too much for that flaw, arriving himself at the other extreme in the end.[34] It is by mapping the intricate reciprocal relation between the two modes that the contemporary philosophy of technology has begun to find a middle road between instrumentalist (and often utopian) and essentialist (and often dystopian) conceptualizations of technology.

Two main characteristics stand out when surveying the approaches discussed above. First, there is the importance of what could be called a shared *ecological* perspective. We can safely say that the contemporary philosophy of technology is *relational* through and through. Again, as we will see, this corresponds with Bateson's framework. That framework, as also already suggested, is about bridging gaps just as much. The instrumentalist and essentialist approaches had and have "subject-object split" written all over them. One cannot expect technology to be either wholly neutral or completely determining if it would not be in one way or another free-standing as object, separated from a subject.[35] But the "empirical turn" perspectives in philosophy of technology take a deliberately relational stance. The matter is not absolutely black-or-white. As we've seen, Heidegger was one of the first to introduce relationality in thinking about tools (although, as noticed by Ihde, his later technology analysis overshadowed that innovative move to a good extent). Surely we can still think about the "essence" of technology and do this in a relational way. Essence is then, simply, relation. If we spread out "determination" widely and thinly enough, we eventually arrive at a picture of technology as indeed determining, but not in a monolithic sense: determination happens then in so many small places, on more modest scales. As Feenberg suggests: "The essence of technology can only be the sum of all the major determinations it exhibits in its various stages of development."[36] (Feenberg is inspired here by Herbert Marcuse, who was in turn inspired by Heidegger.)[37]

Yet if only in order to be able to point this out, the emergence of a relational or ecological paradigm was and is still needed. Tamar Sharon and Mark Coeckelbergh illustrate this in two publications on the human enhancement debate.[38] Both authors aim to foster a breakthrough in that debate, by describing how the human being structurally flows over into domains that were formerly thought to be independent from it—technology, nature—and how the human being thus in a sense disappears; just like, concomitantly, technology disappears. The human being in their view is never some thing, entity, or characteristic that can be effortlessly pinpointed or drawn with just a few pencil strokes. The notion of human nature has become vague and meaningless. Human beings as well as technology are always under construction. When studied from a relational-ecological-holistic angle, the discussion takes on a wholly new appearance. As soon as the terms on which the debate is held have been modified, a beginning can be made with assessing the countless changes that technologies of rich variety bring about day in, day out to our so-called human nature, in other words, our fundamentally dispersed being. Technologies are, just as much, fundamentally dispersed. As Peter-Paul Verbeek observes in *Moralizing Technology*, when discussing the stature of technologies as moral agents, remarking how we don't need to cast technologies as "moral agents *in themselves*": " 'In themselves' entities are quite meaningless anyway—they are given a character in the relations in which they function."[39]

Given this relational bend, it should not come as a surprise that the aforementioned strands in contemporary philosophy of technology line up well with the field of media ecology, and especially with the work of its most prominent source, Marshall McLuhan. The dichotomies outlined above find a pendant in McLuhan's figure-ground distinction. McLuhan was interested in how media and technologies constitute "environments": how they work their effects on our perception, outlook on life, and societal organization.[40] He argued that one cannot understand "mediation" by focusing on content alone. Content in fact mostly acts as a distraction in this regard. For example, television enthralls us with its entertaining and fun programs; all the while we disregard how the medium affects our senses, bodily comportment, way of life, and so on. The latter kinds of impact are rooted not in content, but in the medium's "form"—in McLuhan's rendition: its effects. By adopting a conceptual pair from gestalt psychology, McLuhan seeks to make this difference tangible:[41] content is the "figure" toward which our attention is in the first instance aimed; form makes for the "ground" or background that escapes our immediate attention, but that in fact is of the utmost importance since it *constitutes* the figure and moreover is the place where things "really happen." To study a medium or technology in the proper manner, we need to transcend our short-sighted obsession with content and take a broader view, to observe the medium-as-environment—the medium in its ecological interrelation with countless other media.

Graham Harman, in a paper titled "The McLuhans and Metaphysics," points out the convergence between McLuhan's figure versus ground dynamic and Heidegger's presence-at-hand versus readiness-to-hand dichotomy, respectively.[42] McLuhan himself already presaged this congruence—although he found Heidegger's analysis lacking: "There is in Heidegger still no sense of interplay between figure and ground; the attention has just been shifted from one to the other without trying to take the new thing on its own terms."[43] Still, notwithstanding these supposed shortcomings, a similar dynamic is at play.

This is the dynamic of the central dichotomy. On one hand, in the first instance, we focus on something, put before us, in a purportedly narrow perspective. We see or talk about a "one." On the other hand, subsequently, and with some effort and the guidance of a particular theory and methodology—phenomenological bracketing and variation; McLuhan's project of "understanding media" via the construction of "anti-environments"; but also Latour's and Feenberg's opening of black boxes—we are able to pierce through to a wider view, seeing now that things have the basic characteristic of relationality. Here we see, or now know about, a "many." One mode simplifies, is deficient, incorrect. The other mode reveals the real situation behind the simplification, and that real situation is by definition complex: ecological, relational. Note how Harman also helpfully remarks about Latour (in his other, second book on him, *Bruno Latour: Reassembling the Political*): "Latour often defines 'ecology' as the opposite term to 'modernism.'"[44] In short, there is an intriguing parallelism between all these views with regard to their opposing an ecological point of view to a reductionist stance, with regard to how we deal with technology in everyday situations.

Table 2.1. Four strands of theory and the central dichotomy

	Narrow mode	Wider mode
Phenomenological strand	presence-at-hand	readiness-to-hand
Actor-network theory	closed black box	opened black box
Critical strand	primary instrumentalization / "reason"	secondary instrumentalization / "experience"
Media ecology	content / figure	form / ground

But it doesn't stop there. The difference between the two modes is not merely a matter of one versus many, simple versus complex, deficient versus "full." There is also a difference in *category* between the two. That is the second characteristic. Let me explain. In each case, a difference is at stake between *what is said* and *what is done*. Heidegger's tool analysis demonstrates that we *believe*, on the basis of our metaphysical legacy, that things are "over there": autonomous entities, just as we ourselves are (we assume) autonomously thinking subjects, whereas *actually* what is the case is that all is enveloped in a relational network that is always already "here." With Latour, we find that the Modern Constitution is the way in which we relate the story of subjects/society being separated from objects/nature, while beneath the surface we have in fact never stopped taking part in hybrid realities. Feenberg, in building on this, elaborates his metaphor of the technological unconscious, evoking a tension between on the one hand technically incorporated and on the other hand discursively cultivated values. Either a value is embodied into a technology, or it remains in discourse. Social values that are embodied into a technology disappear into the black box and are not experienced or debated consciously anymore.

We could indeed be seduced into stating that all these dichotomies are instantiations of the classic distinction between discourse and matter, or of discourse and action. There has been in recent decades a "material turn," in philosophy of technology specifically, and in the humanities more generally,[45] calling for a focus on things instead of only on humans. The three discussed strands can be said to be part of this movement (that to a large extent runs in parallel with the empirical turn; for in order to investigate materialities, one has to look at how technologies work and develop in specific use contexts, and vice versa). The material turn certainly often entails a counter-reaction to the decade-long emphasis on language, ideality, and discourse. To illustrate this further, let's look at an approach in which this is particularly pronounced: the work of French philosopher of technology Michel Puech.[46] Puech puts the conceptual pair of discourse-action to work for the study of the existential aspects of living with technology. Actually, his approach constitutes a fundamental critique of discourse in itself.

At the basis of Puech's critique in *Homo sapiens technologicus* lies an observation of a lag of sorts between discourse and action. *Homo sapiens* has become *Homo technologicus*—the human being has become technological—but we aren't able to make sense of that yet, because of an evolutionary incongruence between humans' practical endeavors on the one hand

and our epistemological processes of comprehension on the other hand. Something stands in the way, some latency between the world and our representation of it: our discourse lags behind our actions. Due to that lag, Puech argues, we do not truly realize the potential of digital media like the Internet that could help us in opening up for instance new democratic portals into everyday existential wisdom. Our mindset is still locked in the industrial era. We have not yet adapted to the new situation, to the given that politics, ideology, consumer markets, and advertising as we know them have become completely obsolesced due to the emergence of exactly those new technologies.

We should be able to find new ways of understanding, however. But the change must not take place so much in the domain of *discourse* as in that of *action*. There is a very specific reason for that: technology *is* action. According to Puech, technology is "an action in the real world, not a discourse."[47] Science, on the contrary, is a discourse. This makes technology and science two qualitatively different phenomena. Over the centuries, our prejudices have been skewed in favor of the latter: we have come to prefer discourse over action, "epistemology" over practice. Whereas technological development, says Puech, at any time comes forth not from higher scientific spheres, but from *use*.[48] This crucially means that any problems arising out of that use must be tackled on exactly the same plane, that is, of use. It makes no sense anymore to tackle our technological ailments in purely conceptual ways. "It is outdated to denounce the symbolical parades."[49] Concretely this implies that no longer, Puech states, we are to busy ourselves with attacks on the institutions or on capital—safely abiding within discursive realms. One cannot fight climate change "symbolically" and at the same time retain one's habits of energy overconsumption. We need to undertake action, more precisely, we need to practice ourselves in "micro-actions": small acts of resistance against the archaic forms of dominance. In the context of the example about energy consumption, that could mean—Puech is steeped in virtue ethics—that we change our day-to-day habits, consciously making sure we don't leave the tap running for too long, or minding the temperature of our heating system.

As I said, we could be tempted into literally equating all dichotomies outlined above with the discourse-action or discourse-matter distinction. The issue is a bit more complicated than that; yet there is something important shared by all the strands from the perspective of, call it, a very generic version of the discourse-matter tension. It has to do with that intuitive difference between "talking the talk" and "walking the walk." Contemporary philosophy of technology, generically put, endeavors to challenge the empty posture of merely grasping technology in abstract terms. It tries to go beyond just talking the talk, in order to show what it's like to walk the walk—to show what really happens in myriad contexts in which people technologically *walk the walk*, in which *technologies* walk the walk. And given our first characteristic, this "walking" involves, is constituted of, a network structure. As we've seen with Puech, all this is a way of bypassing epistemological shortcomings. We cannot grasp or understand what technology does, does to us, does to the world, *at first*: we view it as a dumb object, a controllable piece of machinery, a neutral tool. Once again: we think it is *all* that it is, all that we say it is. The central dichotomy serves to get us up to speed in this regard, teaching us it is *not all*; teaching us, almost in a Zen kind of way,[50] to keep silent for a while and let the relational realities stream over us. In this capacity, the central dichotomy is an impressive, massively effective conceptual instrument.

Reflecting on the Central Dichotomy

Nevertheless, notwithstanding its conceptual power, the central dichotomy is also plagued by complications. At least, there are tensions at play within it that deserve some reflection. For one, there is the question of the narrow mode that, as we've seen, casts technology as just efficient means-to-an-end. The central dichotomy subverts the instrumentalist view. Technology can no longer be understood solely from the perspective of the merely efficient realization of goals. This is one of philosophy of technology's revolutionary insights: efficiency is not the only value steering technological development, even though in common-sense discourse we tend to think of it that way. There are other values in play: social, economic, political, cultural, moral values that give shape to technologies. However, as we've seen, those values mostly get forgotten as soon as a technology reaches a certain stage of taken-for-grantedness. From that moment on, we do indeed perceive it as merely efficient (or conversely, as lacking in efficiency). Its multisided genesis gets covered over by the waves of habit and of basic human goal-orientedness. Philosophy of technology has done a great job in uncovering technology's seemingly one-sided façade. Its lesson seems to be: we take technology-as-just-efficient-means for granted, but we shouldn't. There is more to it than efficiency. Or more precisely, technology exceeds our expectation of it to merely fulfill the mandate of efficiency. The net result of this, however, is that this explanation on the basis of efficiency is cast to one side as indeed the purportedly *narrow* view—the perspective to be surpassed by a more complete, *wider* account, namely, of how technology is shaped by so many other factors. That wider account stands over and against the narrow account. And the general idea is that in everyday life, we are naturally disposed to stay stuck in the narrow mode. Only with effort can we go beyond it toward the wider mode in order to see things as they "really are." Once again: this insight has made for meaningful conceptual progress. But it also has had a pernicious side effect, the consequences of which are scarcely investigated: by relegating "efficiency" to one side—the side to be transcended as quickly as possible, for it's only what we think technology is, not what it is—we have gradually become accustomed to taking efficiency for granted. This is probably the reason why Wha-Chul Son, in a volume that studies the legacy of Jacques Ellul, can formulate an initially surprising but ultimately accurate verdict: "Efficiency . . . is a notion that philosophers of technology do not pay much attention to."[51] Exactly, in forcing its breakthroughs of the last decades, contemporary philosophy of technology has steadily been casting out efficiency as a helpful concept with which to understand technology as such—a viewpoint that neatly, and ironically, aligns with its rather skeptical evaluation of "gloomy" efficiency-obsessed Ellul. Efficiency as a denomination and notion has gotten a bad rap.

Not only that, but this is also a stance that makes it difficult, if not impossible, to make sense of everyday situations in which "everything is in its right place": in which all is just running smoothly; when indeed the principle of efficiency seems pretty much in control of things, but not necessarily in the all-permeating, reductive ways exposed by for example Ellul. Here, all simply goes according to plan: no alarms, no surprises, no breakdowns—a situation I would like to designate as "everyday" efficiency. Truly we appreciate efficiency as a highly cherished good. We do this spontaneously. Anyone, for instance, who has ever gone to the airport to take a plane just to find endless waiting rows before the check-in counter can relate to that. We know of course that such hang-ups may have been caused by accidents, bad weather, or random incidents of any kind. But still our whole being cannot help to protest, if only internally, teeth nervously grinding: why can't it go faster? Or in another, less expected context: doesn't one for instance need efficiency to play a musical instrument? Artistic practice is the last place where we would expect efficiency to take center stage. We almost never hear someone say: "How efficient is that guitar player." But still, guitar playing, as anyone who has tried it can confirm from experience, has its *own* efficiency. One has to be able to put one's fingers in just the right position at the right time. This, in a crucial way, is a matter of not wasting resources, namely, of spatial and temporal order: getting your fingers from fret to fret quickly enough. The same goes for, just to name a few areas, cooking, educating children (not to mention getting them to school on time), health, and sports. Even love and relationships have their own efficiency that may require, for example, on a very mundane level the balancing of time spent alone and time spent together.[52]

Philosophy of technology tends to quickly do away with the efficiency notion as just providing the grounds for the "easy" instrumentalist explanation—one side of the central dichotomy: the side to get rid of. As if efficiency is just the icing on the cake, the fancy layer we put on top, in order to decorate the uglier basic structure. But in fact efficiency is all around. Everywhere we look, efficiency influences and steers our activities, albeit not in clear discursive form, but in other, not always recognizable shapes. Efficiency belongs already to the basic structure, to the cake. Everyday efficiency attests to the idea that the efficiency notion deserves a greater role than it now receives in the central dichotomy. This is to suggest that efficiency is in fact spread out across that dichotomy, instead of staying secluded within its discursive safety deposit box on the one side. It crisscrosses both sides. However, we need other means to bring those crisscrossing processes to light; the currently available framework doesn't suffice. At this point, Bateson steps in to help, we will see, by offering an alternative account of *purpose* that doesn't stay limited to narrow purposive rationality alone.

But the wider mode is also not wholly free of issues, in relation to how we engage with it in everyday life. The main question here concerns the extent to which we are capable of—truly, epistemologically—grasping the "networkedness" that typifies the wider mode. Philosophy of technology has fabricated its elegant blend between instrumentalism and determinism, between the definition of technology as merely a means to an end (a "thing") and the view of technology as a substantive realm or system with clear-cut unidirectional effects on humans and society. But this blend has anything but fully found its way into everyday praxis or into the collective consciousness. Here a sharp dichotomy between the two extremes still reigns, and this is illustrated in many recent debates—for example about the everyday use of smartphones, the impact of firearm legislations, the correlation between individual behavior and climate change, the effects of social media on attention spans and cultural attitudes, and many more. We have the hardest time making a connection between our practical-functional use of certain technologies and their broader impacts. Herein probably roots the schizophrenic condition I described in the introduction: we are instrumentalists in everyday technology use—just using our tools and expecting them to be neutral—but determinists when we produce and consume our cultural products—dystopian epics in which narratives of all-devouring technological complexes abound. Thus, in everyday life we find it extremely difficult to see networked realities. Now in a world becoming more radically networked by the day, in which ever more hidden effects seem to go lurking beneath the surface of mere efficiency-orientedness, this issue becomes all the more pertinent.

A crucial problem lies in the given that we have a tendency to choose sides and exaggerate the importance of merely one of the poles: efficiency *or* all-encompassing relationality. At base level, this attests to the idea that *every* dichotomy has its problematic aspects. Actually, the shape of dichotomies as such might be a big part of the problem. Our inclination to always prefer just one side on a two-poled spectrum might be characteristic of all dichotomies;[53] we humans are prone to dualistic thinking, perhaps rooting deeply in our evolutionary history. Imagine being a hunter on the plains looking out for danger: when a predator lurks nearby, it is better to run swiftly into the other direction. This all-or-nothing situation seems to be so ingrained in our worldview that in a world as ours today—having become not only safer in a general sense compared to the hunter-gatherer context, but also much more complex in terms of globe-spanning network constellations—such "digital" thinking, circling around an either-or choice between 0 and 1, has become partly obsolete.[54]

Nevertheless, we need to ask whether, in the case of the central dichotomy, more is going on than this. Is it really merely a matter of epistemology that we prefer to choose either efficiency *or* impact awareness? It does appear that often we can only be in one mode at a time.[55] In order for a technology to work at all, one needs to forget about wider consequences. "Work" is meant here not merely in a functional or technical sense, but also in an existential sense. Imagine what it would be like if you pondered the wider effects of automobile use—traffic jams, pollution, accidents—every time you got into a car. But not just give it a fleeting thought: you would really, by the grace of some magic perception-expanding capability, be able to visualize all of it in one excruciating burst of insight—the stats, the filth, the gore; past, present, and future. To be sure, I would not bet on you never turning on that ignition again. We humans have an amazing gift for blocking out uneasy thoughts. And also, there is the matter of survival: let's say you are terribly ill and need to urgently go to the hospital. You will probably, all nightmarish visions notwithstanding, start that car and drive. But it is not a question of this one time: what if you were *constantly* wrapped up in the wider mode of impact-awareness—would you still drive all that much? So this is a question of the *accessibility* of the network mode in relation to our desire to get things done.

Yet barring for now the purported impossibility, or at least great difficulty, of being in two modes at once, the really interesting question might be this: what happens in the *switching* between modes? This leads us into the matter of the interplay between the two. With regard to the "view-widening" that the representatives of the central dichotomy call for, this issue is important. They call on us to look beyond numb, unconscious tool use, in order to become aware of wider effects—"understand," as McLuhan would have it. For the time being, however, it remains unclear on what epistemological grounds precisely this can be done. If our sight, in the context of our daily doings and technology use, is narrow by structure, how do we make the switch to a wider view? In how far are we actually able to make sense of "networked realities" within a situated, practical-existential context? And how should we then delineate and describe our experience thereof?

A related question—also closely connected to the question of the accessibility of the wider network mode—is this: to what extent is it actually a matter of switching? True enough, we have an almost involuntary tendency to view presence-at-hand and readiness-to-hand in that way. Admittedly, Heidegger himself has somewhat engraved that interpretation into our minds. As suggested, to a large extent it is not even faulty. Of central concern here is the notion of *breakdown*, a kind of "switching point" in the Heideggerian framework between *Zuhandenheit* and *Vorhandenheit*. As long as we are merrily hammering away, the hammer blends into the relational network that we, together with all the things involved, *are*. Safely couched within *Zuhandenheit*, we stay far removed from that nasty metaphysical state of objectifying and picturing. But things suddenly change when breakdown happens: when the hammer gets broken or lost, for example. Then we abruptly (need to) pay attention to the hammer—or the idea of the hammer—in a more conscious way.

Perhaps the pejorative taste of the word "breakdown" plays tricks on our unconscious. But implicitly, Heideggerian breakdown is often seen as a deficient modality: something that needs to be fixed, replaced by something better. Heidegger added insult to injury with his later technology critique. Once again, technology being in this view the ultimate instance of modern subject-object thinking, it represents in fact the extreme, all-penetrating embodiment of presence-at-hand. Everything is objectified in its reign. This analysis does not help in giving *Vorhandenheit* an innocent appearance either, of course. We are left, then, with one mode in which things are running smoothly; another mode in which there is breakdown; the assertion that technology has rather to do with the latter; but also the inherent suggestion, read in Heidegger by later philosophers of technology, that actually what technology is about should be sought with the former, namely, the relational-network-like mode. But could it not be that technology is *not about one mode or the other*, but *really about the connection and interaction between the two*? These two modes, notwithstanding all of Heidegger's own indications, should not be seen as two states between which to switch, but as belonging to each other, like peas and carrots.

Peter-Paul Verbeek already hints at this possibility, although he doesn't nearly go far enough. As I mentioned, the helpful quality of the *Zuhandenheit-Vorhandenheit* distinction is fruitfully demonstrated throughout the explorations and case studies within postphenomenology that in recent decades have followed up on Ihde's pioneering investigations. But postphenomenology, although it sees the distinction as very useful, has also worked to demonstrate the extent to which the distinction is not really, or at least not always, a *distinction*. Especially Verbeek, who perhaps of all postphenomenologists apart from Ihde himself has devoted the most attention to the tool analysis in a deep-conceptual sense, points out how the difference between the two modes is not a matter of a strict split—with a switch in between—as a reading of Heidegger may suggest. In use, technologies might just as much have to be present-at-hand, and not only ready-to-hand. To be sure, the "concept of readiness-to-hand directs our attention to the way in which objects are present in the relation between human beings and their world, and brings such things into precisely the domain that phenomenology investigates."[56] But some tools are not exclusively ready-to-hand when we use them, nor should they be. Verbeek gives the example of a piano: it is partly ready-to-hand and partly present-at-hand.[57] In fact, in order to nourish a durable, sustainable relationship with our artifacts, these might have to be designed to deliberately play upon the present-at-hand, so that we would notice them. The ceramic heater that adorns the cover of Verbeek's *What Things Do* is a case in point: it invites a conscious, continuous engagement instead of working solely in the background, as more conventional contemporary heating systems do. Verbeek's illustrations work excellently in overturning the sometimes implicitly held assumption that technology design and use are only about the ready-to-hand.⁵⁸ The tool analysis and its postphenomenological adaptation show that our dealings with technology always play out on a spectrum from basic, experiential, embodied instrumentality to abstract, cognitive, theoretical reflection.⁵⁹

Nevertheless, we must go further. Even though Verbeek shows how some usage of technology *needs* the present-at-hand, that still does not exhaust the question of the *switch* between ready-to-hand and present-at-hand and vice versa—whether it is possible, and if so, how we do it. The issues I've delineated here point to a persistent unclarity in relation to the interplay between the two modes. Are we capable or not of switching from narrow to wider mode? Are the two modes at the same time there—or here—or not? And after all, is it really about a switch?

My proposal for clearing up these issues is to start using yet other dichotomies, different metaphors, to be overlaid onto the central dichotomy, in order to fully deploy the latter in all its force. In chapters 6 and 7, I will return to these problems—the status of efficiency, the relation between discourse and matter or action, the relation between the two poles of the central dichotomy as such—and reevaluate them from the perspective of those new concepts that I will construct on the basis of the Batesonian framework to be developed next, in Part II. It will become clear that Bateson's work harbors its own versions—albeit implicitly—of instrumentalism, determinism, and some way in between. But precisely acquiring his unusual vocabulary for the philosophy of technology will allow us to get a fresh perspective on their interrelation.

PART II
BATESON AND TECHNOLOGY

CHAPTER 3

Between One and Two

Epistemology/Ontology

Relation is essentially a matter of *two*. One needs two elements to be linked up, combined, united into a new configuration. There's no relation without things being related—that insight seems sufficiently common-sense. But what comes first: the relation or the things being related? Depending on the answer, a completely different ontological outlook emerges.

In the course of philosophical history, the question has indeed been answered in different ways. For a large part of that history, mostly the answer was in favor of *things* coming first. Whether it's in the Platonic theory of pure ideas, the Aristotelian account of all things having an ingrained essence, or the atomism of natural science: the main assumption is that the world is composed of elements that come together to form aggregates. *First* the components, *then* the connection. Only throughout the twentieth century, a relationalist paradigm became predominant, with Alfred North Whitehead's process philosophy, for one, but also with phenomenology, structuralism, poststructuralism, and many other contemporary approaches stressing the primordiality of relation in one way or another. Contemporary philosophy of technology can be said to be part of this stream. As we saw in the last chapter, with the central dichotomy it attempts to retrieve the ecological-relational constellations of impacts and effects beyond the mere appearance of things as objects or efficient means.

Bateson's thought takes part in that dynamic just as much.[1] In this chapter I want to outline the foundations of his relationalism. I will start with elaborating upon that central term with which, as said before, Bateson toward the end of his life purported to describe his main approach: epistemology. For Bateson, epistemology and ontology coincide; what does this mean? And what might it mean for our thinking about technology—more precisely, for how we think technology relates to us, human beings? Subsequently, and largely for the remainder of the chapter, I discuss the epistemology that Bateson wants to put forward. I already sketched this in broad strokes in chapter 1, but here I profoundly go into some of the dichotomies (always dichotomies) that underpin it. We will see how Bateson develops multiple conceptual pairs—structure versus flux, digital versus analog, quality versus quantity, map versus territory—that for a large part overlap, but not always neatly or clearly so. At times, moreover, the correspondence between them seems ambiguous, perhaps owing to the fact that, as also already mentioned, his theory was never completely finished. But can we perhaps create some order after the fact? In a final section I discuss this, and I propose we can. The notion of *purpose* will eventually play an important role in that endeavor; thus we move toward chapter 4.

Epistemology and Ontology

In a fundamental sense, the interplay between "two" and "one" is what Bateson's work is about. We've seen how he acts against materialism as well as idealism (or spiritualism), trying to bridge the gap between mind and matter. But the mistake is to understand those two as *entities*, that then in some way interact. In such a view, either one of them can be seen as in charge of the other. For the Christian worldview and some of its humanist counterparts, mind is the more noble substance, meant to rule matter, that is, the body. In the currently fashionable neuroreductionism, conversely ("we are our brains," with Dick Swaab),[2] matter reigns over "mind." Inasmuch as these approaches still think of things in terms of elements such as *a* "matter" and *a* "mind," that subsequently come together in a certain way, they are instances of a substantivist worldview.

Bateson's approach, by contrast, is avowedly monist and holistic. To get away from substantivism requires a radical shift in thinking. One has to *start* in a completely different way: *not* think of things in terms of elements coming together, but instead assume there is only "one." Of course, we experience not "one"; we experience many things, perceive many things. For Bateson, that sheer fact of perception grounds the reality of life. As we saw, we are dealing here with perception at a very basic level, shared by all living beings, and based in the reception as well as production of differences. That is what epistemology is concerned with. Nonetheless, there is a nuance. Bateson in the end distinguishes between epistemologies (plural) and Epistemology (with a capital "E"). All organisms have or share epistemologies: ways of looking at the world that might be socially constructed and/or biologically determined. But Epistemology (capital "E") is the overarching science that deals with how the world *is* and how living beings cope in it. In terms I will go into more throughout chapter 5: it is situated at least one level of abstraction higher—integrating insights from several disciplines. "Epistemology is an indivisible, integrated metascience whose subject matter is the world of evolution, thought, adaptation, embryology, and genetics—the science of mind in the widest sense of the word."[3] Of course, the two are interconnected: Epistemology is instantiated in epistemologies, and epistemologies take part in Epistemology. This can lead Bateson to the apparent ambivalence of stating at the same time that epistemology is "one, not many"[4] and that "epistemology is always and inevitably *personal*."[5]

This particular dynamic, importantly, leads Bateson to equating epistemology with ontology. A crucial turning point in this regard was the visit he paid to an installation of experiments on optical illusion set up by Adelbert Ames Jr. Ames had Bateson partaking in, for instance, an experiment that reversed the parallax effect.[6] Bateson was shocked by the experience of having his perceptual assumptions completely shaken up. It brought him to the realization that what one perceives—the *product* of perception—and how that perception is brought about—the *process* of perception—are two wholly different things. Of the latter, as Ames's experiments showed, we are scarcely aware. And still those processes help to shape the "product" we are aware of. This is pertinent for epistemology:

> The *processes* of perception are inaccessible; only the *products* are conscious. . . . The two general facts—first, that I am unconscious of the process of making the images which I consciously see and, second, that in these unconscious processes, I use a whole range of presuppositions which become built into the finished image—are, for me, the beginning of empirical epistemology.[7]

There are two main consequences here. One: the more interesting part is what we are not conscious of, that is, not the things we perceive, but how that perception comes about. "What if 'Truth' in some very large and, for us, overriding sense is information not about *what* we perceive (the green leaves, the stones, that voice, that face) but about the *process* of perception?"[8] And two: since these processes are constitutive of all that we experience, *while still unconscious*, then as far as we are concerned, *being is knowing* and vice versa. Epistemology and ontology cannot be distinguished. Mary Catherine Bateson observes: "Because what *is* is identical for all human purposes with what can be known, there can be no clear line between epistemology and ontology."[9]

If epistemology and ontology are one and if epistemology (or more accurately, Epistemology) is centrally about the processes and not the products of perception—*mind* in Bateson's sense—then this has far-reaching implications for how we think about technology, specifically how technology relates to us, human beings. As we've seen, contemporary philosophy of technology assumes the boundaries between the human being and technology to have collapsed—ontologically speaking. The two are continuous. Yet *ontically* we are still faced with decision-making about how far we want to let technology intrude upon us. Bateson will enable us to take a different stance on the problem, just because of his specific version of anti-substantivism, that is, relationalism. Technology is still very "human," in his view, *all too human* in a way. Instead of looking at technology as something that is simply "added" to us (thing) or that besieges us humans—a strange, alien element that comes in to conquer us (system)—we need to understand it as a kind of characteristic that's fully part of our "human personality," but that we can exaggerate and push over its optimum "value" like we can do with mundane characteristics such as kindness, diligence, or tenacity. Interestingly, and paradoxically, the advantages of this approach will manifest themselves as a sort of reinstatement of the split between human and technology—however, not as we know it: "human" and "technology" are simply redefined. Bateson stated of his approach: "I was transcending that line which is sometimes supposed to enclose the human being."[10] It is perhaps for this reason that one can find the odd comparison between his approach and phenomenology here and there.[11] Rollo May sees similarities between Bateson and phenomenologists, but he also points to one main difference: phenomena in Bateson's view "are not perceived 'in general' " as the phenomenologists think, but only when they "differ from something else."[12] Here again the notion of *two* pops up, as well as the ambivalence that characterizes Bateson's epistemology. All is undoubtedly *one*, as far as our conscious awareness of the product of perception is concerned—we are as we know, and we know as we are. Nevertheless, we also always partake in a dynamic of *two*, characteristic of the (unconscious) process that constitutes that perception.

Reality Is Relational . . . but Still Two-Sided

Let's see what Bateson's epistemology-ontology looks like in more detail. I emphasize that what I will be discussing here is Bateson's views as they were taking form toward the end of his life. He was attempting to synthesize the eclectic strands of research he had gathered throughout his career. The results of this work can be found mainly in *Mind and Nature* and *Angels Fear*—the latter book co-authored and finished after his death by Mary Catherine Bateson. Strangely enough, this attempt at synthesis, or at least its conceptual core—his very own *central dichotomy*—is what is often overlooked in readings of Bateson's work.

Recall: Bateson's epistemological outlook concerns living organisms. He is always on the lookout for "the pattern which connects" all living beings. And as we've seen, epistemology works first of all with relations. Making things out of these relations is something that we and other organisms do only "afterward": "In truth, the right way to begin to think about the pattern which connects is to think of it as *primarily* (whatever that means) a dance of interacting parts and only secondarily pegged down by various sorts of physical limits and by those limits which organisms characteristically impose."[13] Mind, Bateson observes, "contains no things."[14] The term *context* serves to make sense of this. Bateson illustrates by way of the hand: we are used to thinking of a (human) hand as consisting of five fingers. But in fact the hand is—from a physiological-developmental perspective to start with—rather a constellation of four relations, between the five fingers. *Context* is precisely this building up of pattern, through time.[15] Living beings are essentially concerned with context. In Bateson's terminology, they are capable of perceiving differences that make a difference. Mind is concerned first and foremost with differences, not with things. "Mental process is always a sequence of interactions *between* parts."[16]

As mentioned, Bateson calls his approach holistic—saying his theory, "like all serious holism, is premised upon the differentiation and interaction of parts."[17] Yet there is a caveat, and it is actually already lurking beneath the surface of what we've just discussed. Not *all* is relation. Reality is not an indistinct pool of naked flux. Indeed, only living things have the capacity of distinguishing differences that make a difference. But this logically implies that nonliving things haven't. And in fact, this is relevant too. It is the difference between the two that carries importance.

Bateson elucidates this by appropriating a dichotomy from Carl Gustav Jung's *Seven Sermons to the Dead*: Creatura versus Pleroma. Creatura and

Pleroma, in Bateson's elaboration, stand for the organic and the nonorganic world, respectively. In both realms, things "happen," but according to a different sort of dynamic. Within Pleroma, there is only *flux*, a mass of indistinguishable events. It is Kant's *Ding an sich*, Bateson suggests, and as such stays out of reach to human (and any other kind of) grasp. Creatura, then, does not stand apart from Pleroma—on the contrary: both are at all times interwoven. But Creatura, unlike Pleroma, works with distinctions. More precisely, it applies them to Pleroma's flux. Creatura inserts discrete marks, gaps, divisions into the continuous flux. It applies *structure*. This is what living things do: they distinguish. Minerals, rocks, planets can only exist within continuity. They cannot perceive or make differences that make a difference.

All this may sound very abstract, but Bateson illustrates by using cybernetics' most famous example: the thermostat. How warm or cold a room is: that is at all times a matter of flux. This is the pleromatic side, where even terms like "warm" or "cold" actually have no meaning (these are already creatural: distinctions). Then there is this technological system in place: the heating system with thermostat. If the temperature (also a creatural term) drops below a certain point, the thermostat will detect this and switch on the furnace. And this is a creatural distinction within Pleroma: a mark within the flux. So it goes on: if the temperature rises to another point, the thermostat will switch off the furnace again. Once more, a discrete mark: a gap in the flux of events.

The example should not mislead us into thinking that this is merely a technical affair. We are talking about a foundational dynamic here, an essential aspect of how the world is structured. Reality is fundamentally two-sided—that is, our reality: the reality of all living beings. Bateson purported to have found with this scheme a viable alternative to the Cartesian composite of matter and mind, of object and subject. The flux-structure interchange transforms the matter-mind dichotomy, shifting the point of difference to the organic-inorganic boundary. *Mind becomes an issue of etching demarcations into the flux of matter.* All living beings are engaged in this. But it's also not a matter of simply reinstating the dichotomy between living and nonliving things—with the former being sentient and active, and the latter just dumb and inactive. It is about the interaction between the two that primordially characterizes the way living beings are in the world; once again, they make distinctions within the flux.

Obviously, saying this, *we as humans* are proclaiming this view. Much in a Kantian style, Bateson to that extent reminds us of our epistemological limitations: "In the end, at the last analysis, everything we say about Pleroma is a matter of speculation."[18] In this sense, again, epistemology and ontology can only be seen as indistinguishable. We know as we are; we are as we know. Nevertheless, what we have found here can be said to already constitute a breakthrough. It lifts epistemology straight out of those thorny discussions about how mind (or spirit) relates to matter, how subjects relate to objects, how words relate to things, how perception relates to so-called objective reality, or how cognition relates to perception. Epistemology "simply" becomes "the science that studies the process of knowing—the interaction of the capacity to respond to differences, on the one hand, with the material world in which those differences somehow originate, on the other. We are concerned then with an *interface* between Pleroma and Creatura."[19] In fact, distinguishing between Creatura and Pleroma becomes the main task of epistemology,[20] and this we do by way of *description*. Description means drawing outlines where there were none. Or more specifically, the outlines are there, but in latent form. They are lying in wait in Pleroma, asking to be *delineated*. "There is a whole slew of regularities out there in Pleroma, unnamed, ready to be picked up."[21] More precisely, because of the fundamental two-sidedness of things, we are concerned with "double description." In fact, it is only by description that Creatura and Pleroma become discernible at all, as Mary Catherine Bateson comments: "Creatura and Pleroma . . . are not in any way separate or separable, except as levels of description."[22] At this point, the true relational scope of Bateson's approach becomes clear.[23] "*Relationship is always a product of double description.*"[24] Put even more strongly, "this double view *is* the relationship."[25]

This two-sided dynamic, the Batesonian central dichotomy, comes back in various corners of his conceptual universe. In order to start making this clear, I discuss in more detail a couple of its variants.

Digital versus Analog

Flux is analog; structure is digital. The analog works with continuous, *more-or-less* differences. The digital is a matter of discrete, *either-or* distinctions.[26] (I follow here Corey Anton, who deliberately distinguishes between "difference" and "distinction" in this sense.)[27] We are acquainted with the terms "digital" and "analog" most commonly from the technical domain: analog and digital technologies. Tape or vinyl records, for instance, are analog: the audio written on them has a form that is "analogous" to the sounds themselves; the grooves of a record correspond physically with the sounds that are produced by them. Digital technology functions in a completely different way: it codes data in sequences of 0s and 1s. There is no physical correspondence anymore between the message and the medium, to put it in McLuhanist terms. If you'd look at a vinyl record with an electron microscope, you could almost "see" the sound. No such thing with a CD: you wouldn't recognize the music that's on it. The code needs to be decoded, converted—by a D/A or digital-to-analog converter—back into (analog) music.

It seems a purely technical affair, a matter for Internet discussion forums about hi-fi, but the distinction between digital and analog in a conceptual sense far surpasses the mere context of engineering. Bateson makes clear how it defines our deepest epistemological orientation. We are constantly involved in making and perceiving digital distinctions into and within continuous streams. Both domains are intertwined, but in a specific way—and this is crucial: the digital needs the analog, because it is *in* the analog that the digital makes a mark. The digital punctuates the analog. In terms to be elaborated in chapter 5: the digital is one level of abstraction higher than the analog. We can illustrate with the example of a jogger: the jogger goes for a run and "feels" her run first and foremost in an analog way; she feels her heart rate increasing, sweat appear, perhaps some exhaustion setting in. Without any means of measurement, however, she cannot put precise numbers on those experiences. If she would track her run, by contrast, through a device such as a heart rate monitor, a FitBit, or an Apple Watch, she would be able to say at a given time that her heart rate is 168 BPM. The change from 167 to 168 BPM is not continuous, at least as far as the device is concerned. It is an either-or choice: digital. Now this may seem not so pertinent—what is more logical and natural than measuring things in this way?—but the difference between digital and analog is not trivial. Which one we stress, which one we adhere most importance to, determines in a wider cultural sense our whole worldview—something that Bateson already foresaw himself, as we will later see.

Quality (or Pattern) versus Quantity

Largely overlapping with the digital-analog dichotomy is the quality-quantity pair. Bateson more often uses the term "pattern," in opposition to quantity. This part of his model may be confusing at times, given that he uses the word "quantity" in a slightly different way than we are used to. Or more precisely, the term has multiple meanings, and Bateson specifically engages with one. We might think of quantity as a specific number. As in: the quantity of containers on the ship (and then we count) is 10,426. But this is exactly *not* the way in which Bateson understands quantity; he employs another meaning of quantity, namely, as an *indefinite* amount. Remember, quantity corresponds with analog: it is a more-or-less (or both-and) affair. We perceive or produce patterns ("outlines") within the quantitative, or *from* the quantitative. Bateson describes the quantitative as the question, and the qualitative as the answer.[28] Quantity asks, "Where?" whereupon quality answers, "There." It is helpful to remind ourselves here of the fact that Bateson is speaking from a perspective—that of Epistemology—that ties together many different disciplines. Often when he discusses quantity and pattern, he does this in the context of issues of biology, evolutionary development, morphogenesis, and so on.[29] For example: "Genotypic explanation commonly invokes the digital and the patterned, whereas the impacts of environment are likely to take the form of quantities, stresses, and the like. We may think, perhaps, of the soma—the developing body—of the individual as the arena where quantity meets pattern."[30] The physical form that you and I acquire is the result of this interchange between the two modes. Among the many possible choices, one is picked: brown hair, not blonde hair; blue eyes, not green eyes. A pattern, each pattern, has to come from somewhere: an amorphous pool of yet-to-be-formed possibility. This may seem abstract—and it is. But at the same time we are talking about a process that is going on around us, with us and in us all the time. Bateson compares it to a chain that breaks under tension: one cannot predict the exact place where the chain will break. "The particular digital answer is provided out of the random. The question, however, comes in from the quantitative, represented by increasing tension."[31] With pattern, something arises, lifting up from that undefined pool: a novelty. But it can only come forth from the pool; "the new can be plucked from nowhere but the random."[32] Somewhere within quantity (flux/analog), pattern is lingering *in potentia*. And with the making of pattern we enter another realm, abruptly, radically: "The qualitative pattern was latent before the quantity had impact on it; and when the pattern changed, the change was sudden and discontinuous."[33] Again, like digital and analog, both are at a different level of abstraction (or "logical type").[34]

Now it gets even more confusing when Bateson equates pattern with *number*. As said, quantity in his view means indefinite quantity. This goes against the grain of our very intuitions when we think about quality, the opposite of quantity. Mostly, we believe, quality has nothing to do with counting. On the contrary, often when the counting—quanti*fication*—starts, and especially when the tendency to count and quantify becomes excessive and obsessive, we fear that quality is lost. Just one example: the measurement obsession and "managementalization" that are now pandemic in health care (e.g., care for the elderly, care for patients in psychiatric hospitals) according to many harbor a threat to the core values of care. When, for instance, contact moments with patients are confined to a specific amount of minutes, we may feel that something valuable is at stake. Care in its qualitative aspects cannot be the object of strict calculation. With Bateson we realize how we are not just concerned here simply with contingent ways of doing, of tackling a practical issue: to count or not to count. These are two completely different *modes* or *views*, reaching deep into our ontological-epistemological setup, that do belong to each other, but that define each other exactly through their mutual disparity. As mentioned, however, he equates pattern—quality—with number, and this might be misleading. Still, as I also discuss later in greater detail, he was already in his day critical of the culture, indeed already apparent back then, of quantification. And this critique is very much continuous with his critique of (scientific) materialism, and of the reduction of all phenomena to quantities (e.g., "energy"). Against that, he stresses vehemently that "quantity can never under any circumstances explain pattern."[35]

So for him, "number," akin to "pattern," has a different connotation than it has in common usage; we should keep that in mind at all times when considering the pattern-quantity pair. Nonetheless, there is the unfortunate consequence that his cultural critique becomes inconsistent to an extent. When considering quality and quantity, he criticizes the monomaniacal focus on quantity. But in the context of the digital-analog pair, he'd rather side with the analog—situating himself thus in a longer lineage of criticism that worries about the possible nefarious effects of the predominance of the digital over the analog. Here is an ambivalence that is never really cleared up in his work. In terms of his cultural critique, however, this is luckily not an absolutely crucial obstacle; it still retains its vigor, notwithstanding the inconsistencies. I will discuss this further in chapter 7.

Map versus Territory

A third important version of the Batesonian central dichotomy, then, is a well-known conceptual twosome as well: map and territory. Here, map is convergent with the digital and with pattern, while territory lines up with quantity and the analog. As we already saw, Bateson's source for this pair is

Alfred Korzybski. Let me take a larger detour through Korzybski's thought, as this will prove useful generally for our framework further down the road.

A somewhat obscure figure in North American intellectual life of the first half of the twentieth century, Korzybski was the founder of the domain of "general semantics."[36] That name is deceptive to the extent that the field is not mainly concerned with linguistic meaning, but with processes of human evaluation as such. "Evaluation" stands here for all reactions, conscious or unconscious, that an organism can have to its environment, and by way of which it tries to give meaning to and make sense of it. Korzybski deploys the notion of "semantic reaction," abbreviated as $s.r$, to denote these processes of meaning-making that are, once again, not strictly linguistic. But something strange is going on with these $s.r$ in our times, according to him (Korzybski lived in the first half of the twentieth century). Grave distortions have arisen in the ways we evaluate, due to our intellectual-conceptual heritage. That heritage is, so he claims, mainly Aristotelian in spirit. Our whole way of thinking got shaped by Aristotelian logic that famously circles around three laws: the law of identity, the law of noncontradiction, and the law of the excluded middle. A is A and not non-A.

We are brought up to perceive the world through this lens of "identification": a thing is what a thing is. Nevertheless, Korzybski points out, writing in the 1930s, recent advances in science have shown that Aristotelian logic is flawed. On the sub-microscopic level, as quantum physics and relativity theory help to indicate, there is only change, flux, energy, a "mad dance of 'electrons.' "[37] It appears from this perspective that a thing is in fact never what it is. It is never identical to itself, for it is always in flux at the most basic level of reality. Yet our everyday, nonscientific $s.r$ are still stuck in the Aristotelian mold: our evaluations do not fit what is actually happening. This, according to Korzybski, creates all kinds of ills, because our abstractions do not correspond with concrete reality. Here he introduces his theory of the stratification of knowledge:[38] human epistemology is essentially layered, and each layer or level makes for an abstraction from the previous one.

At the most fundamental level, there are the sub-microscopic happenings already mentioned. That is the event level. As humans, we never gain immediate access to it. We "perceive" it through our senses and neurophysiological apparatus. Yet that is already a higher level: the object level. Here we see "things," but this perception is only a selection, it is as such an abstraction from the event level. However, the object level is not abstract in the way we usually employ the term: it stands for crude, immediate experience. From there, we go on abstracting in the more common sense of the word. Thus follow the semantic levels, and here we enter the domain of thinking and language. The lowest semantic level is still an abstraction from the object level, or as Korzybski puts it, a higher-order abstraction. But since we can keep on abstracting from abstractions, climbing up the ladder of abstraction, semantic levels can be piled up indefinitely, and we can reach dizzying heights of conceptual generality—as so many philosophies in the past have done.

Korzybski has serious reservations about such all-encompassing abstraction. His point is that our Aristotelian mode of thinking starts with abstract principles—identities, identifications—and then projects these *onto* the lower levels. It goes into the wrong direction. The "natural order" by contrast is from low to high. It starts with the world of events impinging upon our neuroperceptual setup, from which we abstract our objective experience of the world, and then proceeds to form further, higher-order semantic abstractions on the basis of that. Paradoxically, nonetheless, we can only literally *say* something about the event and object levels—the "silent" levels in Korzybski's terms—via abstraction, and when modern science teaches us about the actual state of the sub-microscopic event level, it can only do so by way of abstract concepts and what Korzybski calls "extra-neural means" such as scientific measuring equipment and technology. Be reminded that the event level as such is inaccessible. So it is certainly not the case that we should never use abstractions. Humans are naturally inclined to abstract.[39] However, we should climb up and down the ladders of abstraction in the right way, and that is the Non-Aristotelian way. Most crucially: all hinges on the fact of whether our *s.r* are conscious or unconscious.

That, in the end, is the central goal of Korzybski's framework and its greatest educational aim: to achieve greater consciousness of the process of abstracting. We can do this by training ourselves in Non-Aristotelian thinking, which requires that we do not confuse abstractions with concrete realities. We should not equate the *word* and the *thing*. At this point comes in his famous aphorism: *the map is not the territory*.[40] The word is not the thing it represents. In fact, no word can ever exhaustively grasp what it stands for. Yet in everyday life, Korzybski argues, because of our Aristotelian upbringing we expect that to be the case most of the time; hence pernicious identification mechanisms or "semantic disturbances" such as "magic of words," "hypostatizations," "reifications," "misplaced concreteness," "objectifications," and so on.[41] Such evaluative habits he calls "intensional": we project identity onto what is in reality difference. Certainly in the case of what he calls "allness" terms—e.g., wisdom, justice—the sort of concepts Socrates was so fond of defining in the purest of meanings possible—this can have disastrous "psycho-logical" effects. What we need, instead, is an "extensional" attitude that begins with a realization of difference and retains consciousness of abstracting. Significantly, this necessitates that we become aware of the fact that our epistemology is at any moment self-reflexive. There is always this particular subject, that *I* am, who makes abstractions, forms opinions, holds beliefs, judges.[42] In sum: for Korzybski, the matter rests on a rightful respect for the naturally given hierarchy in levels of abstraction, an awareness of which must be at all times individually practiced.

Bateson, who, as mentioned, was to some extent influenced by Korzybski, shares this concern for accurate abstracting. For Bateson the very act of cognition, of mind, is abstraction. "Our entire mental life is one degree more abstract than the physical world around us."[43] From a flux of processes, the organism abstracts pertinent information. At least at one point, nevertheless, Bateson differs from Korzybski: the former would never insist so much on the "holy" distinction between map and territory that the latter defends, that is, the map is not the territory. Sometimes for all practical purposes the map *is* the territory, Bateson suggests. He thinks humans aren't and shouldn't always be able to distinguish between "the name and the thing named."[44] People, for example, identify with a flag; if someone steps on it, they may get terribly angry. We could easily add more contemporary examples, such as cartoons of prophets eliciting rage. Bateson observes, "There will always and necessarily be a large number of situations in which the response is not guided by the logical distinction between the name and the thing named."[45] We may regret it, and call on the concerned parties to respect Non-Aristotelian logic, however to little avail: word and thing will still be confused. Nonetheless, distinguishing them and getting a clearer view on their interrelation remain good ideals to strive toward.

———•———

Here, for now, the bottom line is the continuity between these three variants of the structure-flux dichotomy. There are still other variations, or extensions of the dichotomy (some of which I will discuss in a later stage): primary

Table 3.1. Bateson's "double description" epistemology

Structure	Flux
Creatura / state / digital / number / pattern / map	Pleroma / event / analog / quantity / territory

and secondary process, calibration and feedback. As said, not all of these variants neatly overlap, and also there are ambiguities and inconsistencies among them. But notice already at this point the general convergence with philosophy of technology's central dichotomy—a convergence I will now, to close off this chapter, examine briefly and preliminarily, in order to investigate it more thoroughly in the next chapter.

The Primordiality of Relation?

One can hardly ignore it: Bateson's interplay between, on the one hand, a pool of unmarked processes and, on the other hand, the making of marks within that flux is very reminiscent of philosophy of technology's tension between relational networks or constellations (as the "true" situation) and a reifying understanding of technology as a well-circumscribed thing (what we think or say it is). Also in Bateson, as in the philosophy of technology, there is a hint of attending to just one side; although with Bateson, we saw, there is ambivalence in this critique.

What is not ambivalent, however, is the primordiality Bateson assigns to relation. "The relationship comes first; it *precedes*," he says.[46] We will find later in more detail how much this is convergent with an approach like postphenomenology, which would state the same thing. This is again to emphasize how Bateson's view is relationalist through and through, and the extent to which he aligns with relationalist and what we have called in contemporary philosophy of technology "ecological" approaches.[47] Relation is what ties the world together, according to this view—it is the building block of reality. Morris Berman, a crucial Bateson interpreter whose work I will discuss more elaborately in the following chapter, describes it thus: "Everything, it seems, is related to everything else."[48] Or in the words of Korzybski, "To 'be' means *to be related*."[49]

Crucially for Bateson, of course—we mustn't forget—relation is a question of *mind*, and mind is a question of *relation*. "Mind is an organizational characteristic, not a separate 'substance.'"[50] We saw already how he eventually states that Epistemology, his overarching multidisciplinary approach is, "concerned . . . with an *interface* between Pleroma and Creatura."[51] Yet as I already suggested, there is a caveat, and it has to do with the "necessary incompleteness of all description, all injunction, and all structure."[52] We saw how for Bateson relation is a matter of "double description," of attending to two sides. Here too, there is some ambiguity (always ambiguity). At times Bateson seems to suggest that description is only about the digital/structure side, *not* about the analog/flux side, when he speaks for instance of the notion of "explanation" as "the *mapping* of description onto tautology."[53] It would appear, then, that description and the act of mapping in this case coincide with the structure side, while "tautology" equates with flux.[54] Nevertheless, in other places he allows for the possibility that description can be done in digital as well as analog ways: "The machinery of description . . . is digital and discontinuous, whereas the variables immanent in the thing to be described are analogic and continuous. If, on the other hand, the method of description is analogic, we shall encounter the circumstance that no quantity can accurately represent any other quantity."[55] Bateson might be referring to two meanings of description throughout his work: one is situated a level of abstraction higher than the other. "Double description" generically refers to the two-sided relational dynamic outlined above. "Description" more specifically, and one level of abstraction lower, denotes the act of naming "things" in Creatura and Pleroma—although we actually cannot say anything about the latter, but still we do so; we cannot but do so. There is paradox here, and in this sense the two meanings could even overlap to some extent.

However it may be, description is always *incomplete*. "However much structure is added, however minutely detailed our specifications, there are always gaps."[56] We will see in chapter 6 how this view, perhaps surprisingly, lines up well with the object-oriented perspective offered by Graham Harman. The gaps arise in structure; the pool of flux has no such gaps. Mary Catherine Bateson comments: "These various kinds of gaps are a characteristic of Creatura. . . . The world of flux does not have gaps in the sense used here."[57] Gaps enable relation. Yet, at the same time, paradoxically, because of these gaps, there cannot be "pure" relationality. The "one" is always interspersed with "twos." And so we will find, oddly enough, *some* kind of "substance" in Bateson's framework, when lining it up with Harman's object-oriented ontology.

Be that how it may, we find a resemblance between Bateson's two-sided dynamic on the one hand and philosophy of technology's two-mode interaction on the other hand. But what is this resemblance worth? Why would Bateson's double description be relevant for us? It may already be clear at this point that beyond the resemblance, the central difference between the two perspectives—Bateson and philosophy of technology—concerns the "standpoint" from which in each case the two sides are viewed. Obviously, in a basic sense, both are theories made by humans (we cannot escape the human stance in this way), and so the viewpoint is always "human." Yet what I am referring to here is the "intra-theoretical" standpoint; the place from which, within each theory, the central dichotomy is considered. For philosophy of technology, this is still the human being. No matter how relational we are invited to become—take into account all those quasi-objects and hybrids, surpass the subject-object split—it is still "us" doing the exercise. With Bateson, this is different. For him the dichotomy spreads out across the living world. Life, that is, *mind*, means: making marks within nonlife, non-mind.

Not despite but *because of* this difference, it will be interesting to overlay the two dichotomies and investigate the consequences of their difference. In Bateson a completely different worldview is at stake than the modern worldview we grew up with, intellectually-historically speaking. This also counts for what philosophy of technology endeavors to do: there, just as much, an attempt is made to transcend typically modern reductions. But Bateson's view might in a sense be the more radical of the two. In order to enable the next step on the path of elucidating that, we need to look into another core Batesonian notion: conscious purpose. More generally, *purpose* is the notion playing the central role here. And purpose might be exactly that which the philosophy of technology has forgotten to a large extent— or phrased more accurately: what it has never really begun to adequately reckon with from the start. This is a blindness, a blind spot, at the heart of contemporary philosophy of technology and at the heart of the central dichotomy, which needs remedying.

CHAPTER 4

Conscious Purpose

As humans, we are gifted with goal-achieving skills, but that in itself is not so special. Finding food, shelter, sex: these are nothing a snow leopard or desert iguana can't do. But something makes us stand apart from the rest of "mind." In our species, the capacity to plan and work toward an aim has been particularly well developed. We envision the grandest schemes—building pyramids, landing on the moon, curing HIV—and set a huge array of forces in motion to realize those aims. Planning behavior has been observed with other species, such as birds and apes, but these haven't been able to cover the planet with gigantic infrastructures and communication networks, to control disease, or to venture out into space.

A taste of this Promethean potency is captured by the notion of the *Anthropocene*, which has recently been stirring a lot of discussion inside and outside the humanities.[1] We can surmise that Bateson would have been at least interested in the notion of the Anthropocene, if the term had been around in his lifetime. In a way, he was a forerunner of the current debate on the dominance of human beings in Earth's ecosystem. As one of those early warners of ecological awareness, he spoke already in the 1960s of a fundamental imbalance. We saw how he figures the world to consist of two sides, interrelated in a specific way: Creatura makes marks into Pleroma. Mind is exactly this capacity to make distinctions within and through matter. All living beings share it. To that extent we humans are nothing special. But then Bateson goes on to sketch a trait that only humans display in large measure: conscious purpose. Conscious purpose is a subset of mind, just a part of it—and according to Bateson a small part at that. However, it has one big nefarious side effect: by focusing too narrowly on conscious purpose, by obsessively and blindly running after our goals and purposes, we lose sight of the larger mind. That larger mind, the wider structure, is purposive in its own way, we will find. But this kind of purposiveness we have lost touch with.

Superposed onto the dichotomy between structure and flux, we will see, is this tension between a narrow view of purpose and a wider understanding of it—in significant ways again reminiscent of the central dichotomy in philosophy of technology. Conscious purpose in Bateson's view is essentially linked with technology. Eventually, a kind of cultural critique will be the end product of his reflections about it. If every theory needs a natural "enemy," an idea one vehemently battles against, conscious purpose would make for Bateson's.

In what follows, I outline and analyze the concept of conscious purpose and the cultural critique connected to it. I start by briefly situating the historical background against which Bateson sees it. Second, I discuss conscious purpose as a notion and its core aspects. Third, I investigate how Bateson sees the relation between conscious purpose and technology. In short: technology is an amplification, an enhancement of conscious purpose—which makes Bateson at least partly convergent with contemporary approaches. Fourth, in order to open up Bateson's framework further to those approaches and broaden the debate, I discuss two other critiques of purpose, namely Morris Berman's analysis and Maturana and Varela's theory of autopoiesis. Fifth, on the basis of that broadened perspective, I delve into the similarities between this "narrow versus wider purpose" dichotomy and the central dichotomy, and begin to elaborate the consequences of Bateson's cultural critique, which will enable us to make an easy jump to chapter 5.

Purpose and Consciousness

Literally and trivially put, the term "conscious purpose" consists of two parts. Both components—purpose and consciousness—stretch back deep into philosophical history.

The notion of *purpose* goes back at least to the Greeks. What is the purpose of a thing, of life? The question has been asked many times over (not necessarily in this exact form), and usually some distinction between kinds of things, and kinds of lives, has come out of the exercise. One of the earliest conceptualizations of purpose is Aristotle's causation theory. Of his four causes, final cause corresponds most with purpose. A thing's or organism's final cause—in other words, purpose—is the aim toward which it naturally strives. Generally speaking, Aristotle attempts with his four-cause model to account for change in the world.[2] As far as the model goes, there is no distinction to be made between natural and artificial objects, except that organisms have their cause of change in themselves, whereas human-made things do not. Yet all things under heaven have a purpose, a "part to play." A similar yet slightly different outlook will subsequently dominate Catholic doctrines. Catholicism pictures the human being and for that matter all other living beings as fulfilling a preordained plan, namely, God's. Here the purposes of all things are, so to speak, infused from the outside. In Aristotle's view, the *eidos* of a thing, its ideal end form, is inherent in the thing itself. Catholicism situates the essence of things—like Plato—in a transcendent principle instead. In creating the world and everything in it, God has given a purpose to it. And although everything that walks, crawls, and slithers the earth is thought to partake in this divine purpose, a special task is set aside for the human being: to guide and lead the world to its final destiny.

Whether one chooses to localize purposes in the things (immanent) or above the things (transcendent), whether one fancies the human species to have a somewhat more exclusive place in the grand plan or not, it remains the case that one is taking for granted the principal idea of linear-historical progression toward certain goals. That idea will begin to crumble under the weight of a new dominant view, with the rise of the scientific paradigm from the sixteenth century onward. The teleological worldview starts to receive blows: no longer are things understood as fulfilling a plan, be it intrinsic or extrinsic. With the methods of empirical observation and repetition of experiments comes the assumption that everything in the world happens because of laws. Not teleological laws: mechanical, physical, "intentless" laws. Purpose falls into disrepute.

Some, like Marshall McLuhan, have evaluated this evolution in terms of Aristotle's four causes, seeing an enthronement of efficient and material causality to the disfavor of final and formal cause.[3] We are well acquainted with all those conceptual pairs that serve in one way or another to connote the fundamental shift in worldview that is at play here: reductionism or mechanism versus holism, right brain versus left brain, C. P. Snow's "two cultures," Gadamer's *Truth and Method*, and so on.[4] Morris Berman employs the well-known notion of "disenchantment," observing how the process involves the decline of what he calls "participating consciousness" to the benefit of a new outlook best described as "nonparticipation."[5] I will return to his account in more detail later on.

Around the middle of the twentieth century, nevertheless, purpose makes a comeback of sorts. It is cybernetics that reinvigorates the notion,[6] after having been declared a no-go zone for serious scientific minds. More precisely, the notion is introduced into the field through a foundational article by Arturo Rosenblueth, Norbert Wiener, and Julian Bigelow.[7] As for Bateson, this article is the origin of the term "purpose" closest to him that he refers to. In their succinct essay, Rosenblueth et alia shed light on purpose from a fresh angle, as they endeavor to make sense of "teleological" behavior. Organisms strive toward goals, and a similar behavior can be observed with certain technological systems, for instance target-seeking missiles. At the time, those were a prime interest of Wiener, who was involved in the wartime effort, doing work on automated aiming and firing of anti-aircraft guns. However, the authors want to explain these "teleological" characteristics across the living and the mechanical world without recourse to determinism or causality. This is no small philosophical task, given the long lineage of thinkers who tried to explain goal-oriented action by invoking some causal principle, such as Aristotle with final cause. Cybernetics' conceptual toolkit, by contrast, in the opinion of Wiener and his colleagues, enables a more "neutral" approach. The notions of self-organization and negative feedback suffice to define teleological, that is, purposive or goal-oriented behavior:[8] purpose in this sense is "just" a matter of self-corrective or negative feedback mechanisms. In reaching for a glass, for example, I unconsciously control my muscles, so that when my hand deviates too much from the preferred path, the error is immediately corrected and my hand steered again in the right direction.[9] Investigating system behavior in this way has a nice practical side effect, in that it becomes possible to predict and control those processes.

Bateson recognizes the breakthrough. He remarks about Rosenblueth et alia's analysis, somewhat drily: "The problem of purpose, unsolved for 2,500 years—came within range of rigorous analysis. It was possible to model even such marvelous sequences as the cat's jump, timed and directed to land where the mouse will be when the cat lands."[10] We will find shortly that he is ultimately much less enthusiastic about purpose than this quote would suggest. Nevertheless, he still embraces the notion as a theoretical tool.

Consciousness, on the other hand, is an altogether different matter. Just as much a long-standing philosophical puzzle, it has at least since the Enlightenment been the subject of specific scrutiny, for example in the work of Descartes and Locke. What is it to be conscious, what is it to be conscious of something, what is awareness? In recent decades, those phenomena have particularly intrigued and busied philosophers of mind and cognitive scientists; with fierce discussions ensuing that I cannot possibly begin to reconstruct here. But in one stream of the field, which is highly influenced by systems thinking, cybernetics, and phenomenology, Bateson's holistic notion of life-as-mind (and mind-as-life) has survived as a kind of background influence.[11] It is also this stream, represented by authors such as Evan Thompson and Francisco Varela, that still systematically clings to the importance of the concept of purpose. This is probably no coincidence. Yet Bateson himself felt hesitant to engage head-on with the notion of consciousness.

Whereas in an earlier work of 1951 he could still aver, though already with sufficient caution, that consciousness "is certainly a special case of codification and reductive simplification of information about certain parts of the wider psychic life,"[12] in later work he remained deliberately vague on the topic.[13] He was concerned that consciousness, like religion, was one of those domains "*where angels fear to tread*":[14] too thorny, too complex to exhaust by way of a simple, linear conceptual treatment. Hence also the title of the book *Angels Fear*. That book does talk about consciousness, but obliquely. Actually, it attempts to understand consciousness *through* the notion of purpose, and in that way somehow delivers an account of consciousness, be it a paradoxical one, where consciousness is rather conceptualized *in absentia*.

Indeed, what is certain to Bateson is that consciousness and purpose essentially relate, as this longer quote from *Steps to an Ecology of Mind* illustrates:

> Consciousness operates in the same way as medicine in its sampling of the events and processes of the body and of what goes on in the total mind. It is organized in terms of purpose. It is a short-cut device to enable you to get quickly at what you want; not to act with maximum wisdom in order to live, but to follow the shortest logical or causal path to get what you next want, which may be dinner; it may be a Beethoven sonata; it may be sex. Above all, it may be money or power.[15]

Already some hint of ominousness is betrayed by this passage. The pact between consciousness and purpose in Bateson's view appears just as much a curse as a blessing. Even in the aforementioned earlier *Communication*, it was already suggested that "many specifically human problems and maladjustments arise from this mirroring [i.e., the codification and simplification of information, referred to in the quote above] of a part of the total psyche in the field of consciousness."[16] Later, Bateson more forcefully states that a question of "perhaps grave importance is whether the information processed through consciousness is adequate and appropriate for the task of human adaptation."[17] Let's go into these issues more thoroughly.

Conscious Purpose

Bateson's account of conscious purpose could be summarized simplistically like this: in mercilessly chasing our goals, we disregard the wider effects of our actions. An image comes to mind of the ever-active entrepreneur go-getter who continuously displays a lot of zeal and diligence—highly praised traits not only in Western but also in some parts of Eastern culture—and who storms toward his target, knocking over and breaking many things in the process, without much noticing. Bateson: "When you narrow down your epistemology and act on the premise 'What interests me is me, or my organization, or my species,' you chop off consideration of other loops of the loop structure."[18] In that process others may suffer, but in the end it will even play to the disadvantage of the involved actors themselves: conscious purpose harbors some ingrained revenge. Like a boomerang, it comes back to hit one in the face.

The reason for this is that conscious purpose, at least potentially, is a matter of maladjustment. It can undermine adaptation. It doesn't *necessarily* do so, although Bateson very much seems to suggest that it does—more often than not. We must remember that he embeds his thinking on matters of epistemology and his notion of mind in an evolutionary perspective. Bodily (somatic) change, and this includes thought and learning (see chapter 5), is similar to evolutionary change, in that it works stochastically:[19] it constitutes a dual process, in which randomly generated variations and external selection processes interact with each other to bring forth "new patterns of adaptiveness."[20] However, Bateson emphasizes, it's a thin line between adaptation and addiction. Many biological processes do seem to stabilize around some "optimum value,"[21] in the sense that, for instance, more of a thing that an organism needs (e.g., food) is not better than less of it. But some of these processes may go awry. "What is good for a short time . . . may be addictive or lethal over a long time."[22] Too much food, as is well known, leads to health issues of obesity and weight-related diseases. A too-successful species may overpopulate, leading to ecological imbalances in an area. What is meant to be a strategy for adaptation becomes a seedbed for addiction, and runaway processes appear, in which positive feedback mechanisms elicit an ever-ongoing amplification of the existing situation.

Straight in between negative and positive feedback, then, conscious purpose does "its work." Clearly it has driven humankind forward in its ruthless development of institutions, technologies, laws, and culture in general. Also, Bateson remarks here and there, purpose is inscribed into our human setup. We are by nature purposive. "I am guided in my perception by *purposes*."[23] There is no nonpurposive human being. But there is also something about our epistemological setup that makes us blind toward its excesses, its potentially pathological development. Comparing us humans to the frog that does not realize it is slowly being boiled, Bateson warns that we have great "difficulty in discriminating between a *slow change* and a *state*,"[24] for example when it comes to the effects of pollution. So if anything can be said about consciousness, it is that it *has its limits*. "How many people are conscious of the astonishing decrease in the number of butterflies in our gardens?"[25] *Ecce homo*, one could say: consciously striving toward all sorts of goals, but shockingly oblivious of the wider consequences of that obsessive focus on achievement.

As mentioned, the part we ignore is, in Bateson's view, significantly larger than the part we—with what is in this context a somehow inappropriate expression—"keep in mind." Conscious purpose deals exclusively with the latter. "Purposive consciousness pulls out, from the total mind, sequences which do not have the loop structure which is characteristic of the whole systemic structure."[26] This is related to the interesting observation Bateson makes every now and then that *purpose saves time*.[27] Purpose is essentially a shortcut: we try to go from A to B to C without taking into consideration all the D's, E's, and F's that are also involved in the web of interactions of which our lineal path from A to B to C is just one tiny slice. In other words, conscious purpose narrows our sight, making us perceive things in terms of simple, one-dimensional causation. Mary Catherine Bateson comments:

> Gregory was proposing a particular and lethal structure to the distortions of perception: that consciousness, shaped by purpose, distorts perception in a specific way, making us think that the world works in lineal sequences. We believe that we can go from *A* to *B* to *C* in achieving our purposes, but in fact each step has multiple effects.[28]

Later she gives the nowadays still very topical example of using pesticides: "The farmer and the pesticide manufacturer, in pursuit of their conscious purposes, have seen a complex interactive system in terms of a single line of causation."[29] But while shortcuts offer benefits—they take us to places faster and with less effort—inevitably at some point one has to pay for them. There is no such thing as a free lunch. The larger network of relations in which our purposive action is enveloped has to balance out in a way. When we cheat *here*, we'll unavoidably have to account for it *there*. Bateson warns: "Lack of systemic wisdom is always punished."[30] And so when it appears that honey bee populations are substantially declining, with pesticides as a key factor in their decline, the boomerang returns hard in our face. It is not just a matter of practical details; "Oh, we just forgot about that aspect; let's go and correct it by tweaking another variable." The deeper epistemological lesson that we should learn is this: "We do not live in the sort of universe in which simple lineal control is possible. Life is not like that."[31]

What we tend to disregard in our unilaterally conscious-purposive disposition is the "whole systemic structure," or with variant phrasings all used by Bateson, "the total network system," "the larger interactive system," the "cybernetic system." He asks militantly: "What happens to the picture of a cybernetic system—an oak wood or an organism—when that picture is selectively drawn to answer only questions of purpose?"[32] And he provides the reply in his typical style:

> If you allow purpose to organize that which comes under your conscious inspection, what you will get is a bag of tricks—some of them very valuable tricks. It is an extraordinary achievement that these tricks have been discovered; all that I don't argue. But still we do not know two-penn'orth, really, about the total network system.[33]

This neglect of the true, that is, "cybernetic nature of self and the world,"[34] or perhaps, more precisely, with Noel Charlton, "the cybernetic connectedness of self and world"[35]—in essence, our inability to see the wider impact and effects of our individual and societal actions—is to be designated as the prime cause of many of our most pressing problems, the ecological crisis to start with.[36] As mentioned, already in the 1960s and 1970s, Bateson was intensely engaged with environmental issues. Referring to some of the problems prevailing in the academic and public debate at the time, such as pollution, atomic fallout, insecticides, and the prospect of the Antarctic ice cap melting, he argued that "this massive aggregation of threats to man and his ecological systems arises out of errors in our habits of thought at deep and partly unconscious levels."[37] To add insult to injury, those problems heighten the sense of emergency and therefore make us even *less* able to act with "systemic wisdom," reinforcing the dynamics of conscious purpose and so actually worsening the situation; I will return to this "paradox of conscious purpose" in the following chapter.

Thus, conscious purpose, otherwise an invaluable asset in human adaptation, at some point has started to impede the process of adaptation.[38] "[T]he world became *addicted* to what was once an *ad hoc* measure and is now known to be a major danger."[39] This nevertheless still raises the question of whether conscious purpose is a general *human* characteristic or whether it is a *culturally* constructed condition. If the world *became* addicted to conscious purpose at one point, that would mean it hadn't been so before. Bateson stays ambivalent in this regard. On one hand, he gives hints about conscious purpose being simply rooted in the human constitution. The phenomena of consciousness and purpose "work" in a certain way, and if they come together, as in humans, specific things start to happen. Consciousness is only a slice of the total mind, but it does have an effect on action, it feedbacks into the total mind, and this has a nontrivial logical consequence:

> If consciousness has feedback upon the remainder of mind . . . and if consciousness deals only with a skewed sample of the events of the total mind, then there must exist a *systematic* (*i.e.*, nonrandom) difference between the conscious views of self and the world, and the true nature of self and the world. Such a difference must distort the processes of adaptation.[40]

Elsewhere, Bateson evokes the metaphor of Adam and Eve. Adam and Eve at a certain point "discovered" conscious purpose, he imagines, began to do "things the planned way," which made them "cast out from the Garden the concept of their own total systemic nature and of its total systemic nature."[41] Subsequently, he observes sardonically, "Adam went on pursuing his purposes and finally invented the free-enterprise system."[42] It may be a metaphor, but of course evoking the Genesis story would suggest that conscious purpose in Bateson's view concerns a deeply fundamental human trait. On the other hand, however, in other places he suggests especially our Western mindset to be oriented toward, or organized in terms of, conscious purpose. In 1977, looking back on the "Conference on the Effects of Conscious Purpose on Human Adaptation" that was held in 1968 (more on this also in chapter 5), he describes conscious purpose as "a sort of fake, an artifact or epiphenomenon, a biproduct [sic] of a disastrous process in the history of occidental thought."[43] This "disastrous process" is definitely concomitant with the distorted lens we put before our eyes in molding ourselves within our modern Enlightenment epistemology—splitting matter from mind, and leaving it at that.

We do not necessarily have to resolve the ambiguity. It is well possible that the human being is generically-structurally at risk to get stuck into conscious purpose's pitfalls, but that at the same time Western thought and intellectual history have heightened the risk for us. We don't even have to restrict this to Western culture. As said, some elements of Eastern culture—if we can use these categories—also align well with the premises of conscious purpose; think for example of the Japanese work ethic. The important takeaway insight here has to do with what happens in a society or cultural setting in which conscious purpose becomes predominant. Whatever the conditions, Bateson seems to say, "If you follow the 'common-sense' dictates of consciousness you become, effectively, greedy and unwise."[44] Or, as Rollo May comments, "Too much emphasis on conscious purpose can lead to a 'monstrous distortion' of integration and a lack of grace."[45] And, importantly, but also tragically, with Bateson again: "The imbalance has gone so far that we cannot trust Nature not to overcorrect."[46] In aiming unilinearly toward specific goals and putting all of our efforts in the service of attaining them, we have lost touch along the way with the gigantic amount of mind that is surrounding us—that is, our shared elementary forms of communication, of exchanging information: the "pattern which connects" us to the rest of nature.[47]

Technology and Conscious Purpose

Now, conscious purpose in itself already poses a challenge. But in our times another factor has been complicating matters some more. In this context—and only here—Bateson specifically comments on technology. Technology, according to him, "added" to the "system," works to strengthen conscious purpose, shaking up that system even more:

> What worries me is the addition of modern technology to the old system. Today the purposes of consciousness are implemented

by more and more effective machinery, transportation systems, airplanes, weaponry, medicine, pesticides, and so forth. Conscious purpose is now empowered to upset the balances of the body, of society, and of the biological world around us.[48]

Conscious purpose gets "implemented" in and thereby "empowered" by technology. It is a kind of extension or amplification of it; an embodiment or a materialization, if you will, as far as technology can still be regarded as purely material—but we have to remember that Bateson is writing in the 1960s and 1970s, a time in which digital technology was still largely in its infancy. Notwithstanding possible conceptual subtleties with regard to the difference between the two terms, in the following statement "technology" could very well have been substituted for "design": "I suppose 'design' to be the physical realization, first on the drawing board and then in metal, of conscious purpose."[49] Technology instantiates conscious purpose.

The result of this is that technology thus substantially enhances and worsens the problems brought forth by conscious purpose: "It is in our power, with our technology, to create insanity in the larger system of which we are parts."[50] In general, Bateson perceives modern industrial society as a textbook example of a runaway process,[51] in which values such as profit-making, quantification, control, and consumption are excessively prized and stepped up to pathological levels. Somewhat similarly to McLuhan, Bateson feels we might be moving along too quickly with the developments made possible by technological advances. While McLuhan advises to "think things out before we put them out,"[52] Bateson observes (albeit rather as a side note to a broader discussion) that we should not automatically assume that the *new* is always better than the *old*—referring to technologies such as "electronic communication devices." We shouldn't trust "blind stochastic processes" to "always work together for good," he says, implicitly viewing technological development here from an evolutionary perspective.[53] The new might not even be viable, while the old has demonstrably proven its worth. With some qualification: "Other things being equal (which is not often the case), the old, which has been somewhat tested, is more likely to be viable than the new, which has not been tested at all."[54]

Yet Bateson is no Luddite. He emphasizes that we should not return to some technology-less state, if that were even possible. That would just restart the "whole process."[55] As Mary Catherine Bateson observes, humans cannot help developing "systems of knowledge," and as time proceeds, these grow at an exponential rate:

> There is no way for Western civilization to move backwards . . . returning to an ancient stability. Even if we manage to escape our addiction to economic growth and a continually rising material standard of living, the search for knowledge is embedded in our civilization, and systems of knowledge grow exponentially as each new fragment interlocks with the whole.[56]

But Bateson goes further. It would appear from the above that he perceives technology as a necessary evil, the effects of which we just have to mitigate as much as we can. Yet he also points at its usefulness—if deployed in the right way. In reference to the "systemic wisdom" so badly needed for the remediation of conscious purpose, we find him suggesting the following: "A 'high' civilization should . . . be presumed to have, on the technological side, whatever gadgets are necessary to promote, maintain (and even increase) wisdom of this general sort. This may well include computers and complex communication devices."[57] Moreover, in line with the paradoxical character of conscious purpose, technological progress—be it that the term "progress" sounds cynical in this regard—is cast by him as inevitable.[58]

We can already partially see how Bateson would line up with the philosophy of technology. Anticipating a more extensive discussion of the convergences in a later section, let me give some quick pointers here. As regards his definition of technology—implicit as it remains—he is most akin to McLuhan's (and others') concept of technology as an extension of the human body, senses, or capabilities. Considering his theory of mind, moreover, his analysis can be seen to foreshadow in a way the "extended mind thesis," which delineates technology as a full-blown part of cognition.[59] Problematic nevertheless are his suggestions that technology is "added" to existing societal or biological systems, and that we somehow could hold up "the new" to clear objective circumspection. Current perspectives in philosophy of technology frame technologies as constructed *in* and *through* societal contexts, pointing to the faultiness of that image, that is, of seeing technology as an addition, something that is brought in (or "put out," in McLuhan's terms). At the same time, obviously, Bateson qualifies and nuances his observations all the time. At one moment, again in line with the extended mind thesis, he suggests that tools are capable of being part of mind, hinting at "larger circuit systems extending beyond the limits of the skin."[60] And we can find Mary Catherine Bateson elaborating upon her father's theory by way of the example of a peasant wielding a scythe, asserting that tools are just as much a part of mind as the so-called natural elements: "Mind is immanent in the system man plus scythe."[61] In order to straighten out these issues and further crystalize the Batesonian philosophy of technology, I want to expand the discussion a bit and take on board two critiques of purpose, essentially related to Bateson's, that can help us to fully unpack these ideas.

The Critique of Purpose: Autopoiesis and Participating Consciousness

Even though Bateson fervently demonstrates the perversities of conscious purpose, it might not be immediately clear to every reader what precisely is so *wrong* with it. One may actually feel that conscious purpose as a typically human characteristic is something we should be proud of: a trait we should cherish instead of denounce. And why not? Indeed, one may state, from the same abstract point of view that Bateson uses, that conscious purpose is also what made us humans do great and admirable things: build pyramids, contain polio, fly to the moon, and so on.[62] So let's probe a little further. Specifically, I believe we should (1) dive more deeply into epistemological considerations and (2) explore more broadly overarching cultural circumstances. I do this by way of Maturana and Varela's concept of autopoiesis and Berman's cultural-historical Bateson reading, respectively. Both are also critiques of purpose, generally speaking, but they highlight different aspects that we need to take into consideration.

While Bateson was gradually systematizing his theory on the ecology of mind during the 1970s, a pair of biologists in Chile were starting work on much the same issues. Humberto Maturana and Francisco Varela eventually arrived at a theory of mind, too, now known as the Santiago Theory of Cognition,[63] that sees cognition as pertaining to all living things. Their framework starts out as a critique of terms usually used to explain life, *purpose* being one of them. This might seem strange, given that we've seen how the notion of purpose had long been cast out as explanatory principle, only to return in the 1940s within cybernetics, a field Maturana was moreover influenced by. Yet what Maturana and Varela rail against is a very specific interpretation of purpose, such as that employed in approaches defining living systems in terms of function and goal. The clichéd reading of evolutionary theory offers one example: living beings are looked at as "achieving the goal" or as "having the function" of adapting to an environment. Maturana and Varela want to discard these notions from the beginning. One cannot characterize living systems as fulfilling some kind of purpose, being adapted to a specific environment or, if one insists, carrying out some higher plan of divine creation. Those are *external* explanations: they account for living systems by reference to an externality. To achieve conceptual rigor we need an *internal* definition of life.

At this point Maturana and Varela come up with the notion that will probably be forever connected to their names: *autopoiesis*. Autopoiesis, or self-production, is the defining mark of the living. A living system is a system that "continuously generates and specifies its own organization through its operation as a system of production of its own components."[64] This process makes it a unity. And that unity has a large amount of autonomy, so it appears. Maturana and Varela elaborate this in terms of determination and the mutual influence between organism and environment. Contrary to what we tacitly tend to assume in thinking about adaptation and learning, they assert that the environment does not determine what occurs in a living being. It can only trigger changes. The structure of the living system itself, by contrast, determines what happens to it. Only its own autopoietic organization has determinative leverage, so to speak. This seems to go against the grain of a good deal of our unstated suppositions concerning the interactions between living beings and their environment. Maturana and Varela redefine these interactions as "structural couplings":[65] "mere" compatibilities or congruences between the structure of the living system and that of the environment. The latter produces "only" perturbations. Beyond that, the autopoietic unity does what it does, because of its autopoiesis. There is a great degree of tautology here; but that is the whole idea.

The dichotomy is not only ontological, it is just as much epistemological. Or put differently, in order to understand life in this way, we need to take the right viewpoint first. That viewpoint is the internal stance. Maturana in the introduction to *Autopoiesis and Cognition* uses the terms "phenomenal domain" versus "domain of descriptions" to elucidate this.[66] The difference between the two depends on the standpoint we take as an observer. If we look at a system from outside, we are in the domain of descriptions. That is the viewpoint we take when we define a system in terms of purpose or function. But as said, this is a skewed view of what actually happens in an autopoietic system. One needs to define the system "from inside." That is the phenomenal domain. In *The Tree of Knowledge*, Maturana and Varela offer a crucial illustration of this matter.[67] They ask the reader to imagine someone who has lived in a submarine all his life. We, observers, stand on the shore and see how the submarine surfaces, skillfully evading the reefs. We switch on the radio to congratulate the navigator for his submarine-steering skills. The navigator, however, is baffled, and answers: "What's this about reefs and surfacing? All I did was push some levers and turn knobs."[68] For the steersman, there is no purpose of reef-evading. There is only the implicit, unconscious self-reproduction of the system of which he forms a part. And this, let us be reminded, is cognition. Whatever characteristic an outside describing observer wants to further assign to the system, from the perspective of the phenomenal domain it is simply irrelevant.

It might appear at first glance that Maturana and Varela in this way have banished purpose altogether. If there is such a thing as conscious purpose, in the Batesonian sense, we surely cannot use it to make sense of what human life is. Yet it is not because they ban the notion of purpose as an explanation of life that purpose has absolutely no role to play as such. On the contrary, as later authors claim, Maturana and Varela might have been too quick in rejecting the purpose notion altogether.[69] Organisms might be very well understood in terms of purpose; they strive toward goals, develop activities and behavior in the service of goals—why couldn't we understand that as a purposive dynamic? That is, to be clear, *without* falling back into purely, old-school teleological explanations in terms of essentially ingrained or "given from above" purposes. In fact, Varela himself changed his mind afterward. He and other scholars on the outskirts of biology, cognitive science, and physics, such as Stuart Kauffman and Terrence Deacon, have in recent years tried to (re)integrate purpose into their disciplines, often on a more general and not strictly human plane, but always in an attempt to correct what they perceive as the overly reductionist bend of natural science. All regard purpose—remember, the term long vilified in the realm of orthodox science-making—as a central notion that in one way or another represents a missing piece of the puzzle. Andreas Weber and Varela reinterpret Kant, who is usually taken to be a forerunner of the reductionist paradigm, as actually anticipating the idea of autopoiesis, with his notion of "natural purpose," which holds that a thing, that is, an organism, is both cause and effect of itself, making it a self-organizing being.[70] By adopting the principle, Weber and Varela seek to reimport teleology into modern biology. Stuart Kauffman, then, forewarns against reducing biology to physics, pointing out how biology needs to account for agency, and with that for meaning, values, and purpose, and these cannot be cast in terms exclusively derived from physics.[71] And Terrence Deacon, most recently, endeavors to track down the "relationships" that the materialist-reductionist approach cannot "see" and that in fact exist only as a sort of absence—Deacon therefore dubs them "absential"—notwithstanding that they do have a perceivable effect upon the world.[72] These include most characteristically purpose but also function, meaning, intention, significance, and value.

Now let us return to the submarine metaphor. Even though we choose not to take the external point of view—the domain of descriptions—we can still understand the internal happenings of the submarine—the phenomenal domain—in terms of purpose. But it is a *wider* sense of purpose. By way of description, we arrive only at *narrow* purpose: I see a thing *x* move from A to B in a certain way, and so I conclude that this is an outcome of "its" purpose. Yet this is a stance that disregards the myriad factors that had to be in place for this thing *x* to even develop what appears like purposive behavior. We bump here again into the distinction between *what is said* and *what is done*. What is said is just a simplistic summary; a slice of a much bigger whole. What is done envelops, takes part in the loop structure—as Bateson would phrase it—flowing over into the larger "mind." The lesson learned from these newer perspectives on purpose could be that, sometime in the past, we got our signals crossed with regard to the true nature of purpose, and we should now reevaluate narrow purpose in relation to (rediscovered types of) wider purpose, and vice versa. Admittedly, the question remains to what extent the concept of autopoiesis can be expanded to include nonliving, "social" systems. Maturana and Varela themselves retained reservations at the time,[73] although cyberneticlan Stafford Beer, who delivered the preface to the second part of their *Autopoiesis and Cognition*, enthusiastically proposed it can.[74] Niklas Luhmann, as is well known, has elaborated the autopoiesis notion effectively in social theory.[75] And Bateson, also, would probably advocate a broader interpretation, as was suggested above.[76]

What this means is that we have not given up on wider purpose—even though it might not necessarily be of the old-fashioned grand cosmic variety anymore. We humans still *need more* than merely narrow purpose. Of course, this is also what the notion of disenchantment is meant to convey, namely, that although means-end rationality or purposive rationality predominates, we remain at our core "larger meaning"–seeking beings. But we arrived at a situation or in a time period in which apparently we are more inclined to take the external perspective and explain things in terms of narrow purpose. How did we get here? And how do those two kinds of purpose relate? At this point, Berman's insights on the "reenchantment of the world" can be of aid.

With the emergence of the natural-scientific paradigm, as we saw, purpose is relegated to the backburner of intellectual history. This shift has often been described in terms of the disenchantment of the world. Morris Berman employs the notion, observing how the process involves the decline of what he calls "participating consciousness" to the benefit of a new outlook best described as "nonparticipation."[77] Participating consciousness entails the interwovenness of observer and observed. It assumes that not only the observer has "mind," but that the surrounding world does too (and actually seen from this perspective, the idea of a world being "surrounding" makes no sense to begin with). Mind infuses everything. Like Bateson, who is his main source, Berman is not suggesting a form of panpsychism. The image of the pre-scientific outlook that he outlines bears resemblance to the brief sketch of a ubiquitously purpose-invested world—be it Aristotelian, Catholic, or anything else—that I previously provided. The notion of mind exactly means to convey a sense of that purposive directedness, the teleological plan in which all things find their "reason" for being (although Bateson would not side with either Aristotelianism or Catholicism). By contrast, the scientific stance largely disposes of this view, replacing it with a mechanistic materialism. Berman comments: "The universe, once seen as alive, possessing its own goals and purposes, is now a collection of inert matter, hurrying around endlessly and meaninglessly, as Alfred North Whitehead put it."[78]

Yet it's not that purpose altogether disappears. Another kind of purposiveness comes to the fore. Berman distinguishes between *immanent purpose* and *purposive manipulation*. The world is no longer understood as pervaded with its own inherent purposes; what remains is mere utility, means-end rationality. With a nod to Weber, Berman argues: "Once natural processes are stripped of immanent purpose, there is really nothing left in objects but their value for something, or someone, else. Max Weber called this attitude of mind *zweckrational*, that is, purposively rational, or instrumentally rational."[79] With this emerges the Cartesian or technological paradigm, which equates "truth with utility, with the purposive manipulation of the environment"[80] and, importantly, *cognition with technology*.[81] Consciousness is no longer viewed as participating, partaking in the larger mind. It has become nonparticipating, exclusively set to control the world that was once thought to be suffused with a power greater than us human beings:

> Things do not possess purpose . . . but only behavior, which can (and must) be described in an atomistic, mechanical, and quantitative way. As a result, our relationship to nature is fundamentally altered. Unlike medieval man, whose relationship with nature was seen as being reciprocal, modern man . . . sees

himself as having the ability to control and dominate nature, to use it for his own purposes. Medieval man was given a purposeful position in the universe; it did not require an act of will on his part. Modern man, on the other hand, is enjoined to find his own purposes.[82]

Hence, in the scientific worldview, in the Cartesian paradigm, the human being comes to be the sole purveyor of purpose, the only entity remaining capable of envisaging purposes. Purpose used to be situated on this side of the organic-inorganic divide, meaning that the whole of the living world shared in it. True enough, we human beings occupied a special position, as in the Catholic perspective, and to a certain extent already in the Greek account (the human being is mostly seen there as the only living being capable of true reason). But at least to a large extent all the living world was considered as coherent, integrated through the principle of common directedness. Now the boundary line has relocated more radically. With the Cartesian worldview's introduction of the infamous subject-object split, purpose is being restricted to human nature alone, as the deliberate, intentional, planned act of controlling and manipulating—in other words, conscious purpose. Not only are we the chief purpose-distributing agent, we withhold it from any other thing, expecting that thing to just become an element within our purpose, thus, indeed, imposing narrow purpose on the rest of the world. And this, if we care to follow Berman's suggestion, is still very much the world we live in today.

A couple of elements are particularly worth noticing in this account. With nonparticipation, Berman observes, the rigid distinction between observer and observed arises. We can see a parallel here with Maturana and Varela's call for an internal approach; indeed, looking out from inside the submarine, one tries to take a participating stance. But they also diverge: Berman sees technology as a prime characteristic of the nonparticipation paradigm. In fact, as mentioned, technology equals cognition—an identification begun by Francis Bacon, who saw technology as "the source of a new epistemology."[83] This marks "the elevation of technology to the level of a philosophy,"[84] embodied in the notion of the empirical experiment. Berman remarks by way of Oskar Kokoschka: "To know something is to control it, a mode of cognition that led Oskar Kokoschka to observe that by the twentieth century, reason had been reduced to mere function. This identification, in effect, renders all things meaningless, except insofar as they are profitable or expedient."[85] But whereas Maturana and Varela would criticize the notion of function too, they would probably not identify technology with cognition straight out. They might see technology as partaking in wider purpose just as much. Or more precisely, extending and elaborating upon their theory, and learning especially from their submarine metaphor, *we should see technology as such*. Of course, in their framework, the illustration is a *metaphor*; but what if we would take it *literally*? What if the *inside* view of technology, of using a technology, can perfectly be described in terms of wider purpose? Is technology a matter of narrow *or* wider purpose then—or both?—and how does Bateson's conscious purpose notion figure in this scheme? This brings me to the concluding section of this chapter.

Bateson and the Central Dichotomy

It should already to a large extent be clear how Bateson's framework dovetails with the approaches to technology delineated in chapter 2, and with the central dichotomy that these approaches share between narrower, efficiency-oriented and wider, impact-aware modes. Conscious purpose can be compared to instrumental reason, Weber's *Zweckrationalität*—the principle that *drives* the development of technology but does not *exhaust* it, as other, social-political values get incorporated into it (Latour, Feenberg), unforeseen side effects arise (McLuhan), and in any case our epistemological-phenomenological makeup is prone to exclude certain factors when considering technology (the Heideggerian tradition and postphenomenology). In Bateson, specifically, the "other side" of purely instrumentally oriented technology use concerns the "cybernetic system," the cybernetic relation between self and world, the network structure.

Bateson's cultural critique consists in denouncing the importance we attach to conscious purpose—even leaving apart the question of whether conscious purpose is a humanly ingrained characteristic or a culturally acquired trait—and instead directing our sight toward the wider mode. In a sense, he shares the mission of philosophy of technology: to rebuke instrumentalist as well as determinist perspectives. Technology is an extension of conscious purpose and in that way, Bateson does seem to nourish an instrumentalist understanding of technology. Looked at in isolation, this idea of technology extending goal-orientation indeed lines up with instrumentalism (and we will see in the next chapter how in a similar way there is also a version of determinism to be found in Bateson's thought, to the extent that conscious purpose, strengthened by technology, is seen somewhat as "the root of all evil"). Nevertheless, when he frames technology as potentially playing a part in the obtainment of systemic wisdom, or when he hints at the possibility of technology being just as much a part of mind, we see an approach reminiscent of the middle-road perspective of contemporary philosophy of technology. Our excursion into autopoiesis theory and Berman's cultural history helped to underscore this ambiguity. Yes, historically and culturally speaking, we exhibit this proclivity toward the technology-as-control paradigm, but such should not preclude technology to be understood also as a wider-purpose phenomenon. In fact, we would do well in trying to frame and understand technology in that way—lest we stay stuck in the impossible choice of *either* just regarding technology as efficient means-to-an-end *or* trying to map the network constellations making up technology but in the process losing our object of study.

But to really probe how far that cultural critique can bring us, we also need to inquire into what Bateson proposes *to do about* it. Such would be our natural impulse: we see a problem and think immediately, "How can we fix this?" But matters are far from simple in this regard. One does not simply fix conscious purpose. Setting ourselves the *goal* of remediating conscious purpose won't do. We cannot just notice the effects of conscious purpose, of obsessive planning, of our tendency to intervene, and then draft up yet another plan, another intervention, another goal to achieve. Because, logically speaking, this merely consolidates and extends the conscious-purposive mode, resulting in still other side effects. How to escape this vicious circle? We will see how in Bateson's view conscious purpose—and our distorted evaluation of it—is a matter of levels of abstraction, and more precisely, of confusing them. So our task would be first to try and get a grasp on these, and subsequently to learn to navigate them.

CHAPTER 5

Remediating Conscious Purpose

How do you rid yourself of a pernicious but unstoppable inclination to get things done, to achieve aims, to set targets, and run after them as if life itself depended on it? Can you will that disposition into nonexisting? Can you undertake specific actions "in order to," or work out a strategy, draw up a battle plan, to achieve the aim of not achieving aims? Once again we find paradox. Bateson himself was all too well aware of this. Is there a way out of conscious purpose *without falling back into conscious purpose*? Throughout his work he seems to have been mainly pessimistic about the conundrum, but eventually he started to see some light at the end of the tunnel—a solution, but not as we know it: no "solution" at all in fact, more of a way to recast the problem, based in exactly the nature of paradox. Specifically, the way in which paradoxes come about can give us hints on how to step out of them.

The key are levels of abstraction or "logical types." Conscious purpose has to do essentially with the confusing of logical types. We already found the idea of levels of abstraction stashed within chapter 3: Creatura (map, structure, digital) is one level of abstraction higher than Pleroma (territory, flux, analog). But this would suggest that there is one basic gap in the world: between the level of Creatura and that of Pleroma. In actual fact, levels of abstraction pervade life and the world beyond that one big gap. We could look at them as a third dimension of sorts, after the two dimensions we already met in the previous two chapters, of structure versus flux and of conscious purpose versus ecological awareness. This one, this third dimension, is in the first instance about perception: awareness of contexts. Again, we need the right epistemological filters placed before our eyes before a breakthrough in any other sense can be made. Art, preeminently, helps with this.

I will first trace the contours of the paradox of conscious purpose. We cannot consciously-purposively reason or act ourselves out of the clutches of conscious purpose. Because we then still remain *in* conscious purpose—like believing you woke from a dream while actually you're still sleeping and dreaming, having only awakened from a dream *within* a dream, as in the movie *Inception* (2010). There are reasons for pessimism in this regard, for Bateson. Yet at some point he began to believe that there might exist remedies as well. I will briefly sketch that turnaround in his thinking in a subsequent section. Then I will go into the specifics of the "solution" he envisions, in two steps. First, I will discuss the ideas of abstracting and of levels of abstraction. Second, I will outline the ways in which Bateson sees the navigation of these levels concretely possible—a process epitomized in his notion of "learning." Art is one of the most important forms in which this process can appear. Thus, finally, I will elaborate on Bateson's views in relation to art, and synthesize the insights we will have gathered up until that point throughout this chapter.

The Paradox of Conscious Purpose

So, what to *do* with conscious purpose? Bateson often seems to suggest there is a kind of wisdom possible, an awareness of the wider network system, "recognition of and guidance by a knowledge of the total systemic creature."[1] If this sort of wisdom did not exist as at least a possibility, Bateson's very own work, bent as it is on mapping the cybernetic system, would be torn asunder, logically speaking. Therefore, there *must* be a way.

Charlton comments that the wisdom concerns "the use of our wider, deeper, more-than-conscious minds," "some forms of wisdom that the other animals still have."[2] Indeed, at times Bateson uses the conceptual pair, derived from psychoanalysis, of primary and secondary process as another placeholder for his fundamental twofold of flux and structure—but also for the dichotomy between systemic wisdom and conscious purpose.[3] Primary process is all about wish fulfillment. In the id, which is wholly unconscious, conflicting drives can exist right next to each other. Or more precisely, they are not distinguished to begin with. This is the prelogical realm; dreams are its main manifestation. Secondary process, then, stands for conscious, logical thought and reality testing, functioning through the ego. Here, there is a need to solve the contradictions bursting through from out of the flux of the unconscious; one *has* to make a choice. You cannot both hate and love your mother at the same time, says your ego, so you have to choose—a conflict that then might be resolved, in the psychoanalytic framework, by way of defense mechanisms such as repression, neurotic symptoms, and what not. Systemic wisdom, according to Bateson, requires a turn away from our conscious, rational (and purposive) thinking, toward the deeper layer of primary process, that by its nature is much more in touch with relational constellations.

Still then, there appears to be a crucial problem; for how to achieve such wise awareness? Can we consciously aim at it, strive toward it? In that case, as already suggested, we are right back in the realm of conscious purpose. Douglas Flemons in his book on Bateson and Taoism sketches the matter exquisitely:

> One who attempts simply to "fix" a systemic situation, be it biological, societal, or familial, is poised from the outset to commit ecological blunders. When action is organized as a solution to a perceived or believed deficit, it necessarily becomes defined in opposition to that which is identified as a problem. . . . Gregory Bateson . . . explains that any action stemming from the desire to do something—indeed, anything—as a corrective of "pathology" is fraught with difficulties.[4]

How can we "solve" the problem of conscious purpose? As mentioned, this seemingly natural way of framing the issue in terms of problems and solutions already leads us in the wrong direction. Bateson displayed an almost fatalistic attitude in this regard. Mary Catherine Bateson in a memoir of her father remarks: "He had all his life the belief that efforts to solve problems are likely to make them worse."[5]

The question seems to be to what extent conscious purpose determines our room for maneuver: how much of human thinking and doing is led by it? Pretty much, it would appear. To the degree that conscious purpose is "chiefly concerned with getting from point A to point B in the most direct way"[6]—"making us think," as Mary Catherine Bateson adds, "that the world works in lineal sequences"[7]—we must concede that it is relatively ubiquitous. A good part of human activity is centered around what could be dubbed this "A-to-B principle." We should thus expect to have the greatest trouble in even trying to *think* otherwise, let alone bypass it in practice. We humans, at least if one cares to follow Bateson, are overly primed and trained in dividing up the world in parts and fragments, ordering and categorizing these parts and eventually manipulating them in the service of well-circumscribed future goals. Trying deliberately *not* to do so entails the pursuit of the exact same tactics. As Bateson puts it in an interview with Stewart Brand: "The moment you want to ask the question, 'What do you do about it?,' that question itself chops the total ecology."[8] What shows here is the essentially paradoxical nature of conscious purpose—or at the very least, of conscious purpose's consequences. And, as we saw in the last chapter, technology makes things worse, amplifying as it does conscious-purposive modes of thinking and doing.

Let us abide a bit longer with the commentary by Flemons, who has perhaps most poignantly identified the paradox. He demonstrates how it has to do with the way in which conscious purpose is structured, is built. Conscious purpose is forced, like secondary process, by the need to make choices. It is based in dualism, unlike the flux of primary process, from which dualistic oppositions are absent. Again, with regard to perceiving things in terms of problems and solutions, Flemons observes:

> It is very difficult to do anything that does not simply make matters worse. Solutions that do not escape the dualistic assumptions of conscious knowing will themselves become symptomatic expressions of purpose, hopelessly entangled in the problems they are helplessly trying to eradicate. In fact, it is precisely this attempt to *eradicate problems* that secures the failure of dichotomous solutions and guarantees an exacerbation of the original situation. Solution and problem are mutually defined; it is impossible for one side of a distinction to destroy the other side—*any* oppositional response only highlights the relationship, thus intensifying the problem and reiterating its intractability.[9]

Consciousness cannot help but to "act" in this way. It "knows by dissecting and analyzing, by *separating*."[10] It can only grasp one side of a distinction, not both (hence the title of Flemons's book, which refers to the eventual ideal of overcoming this limitation: *Completing Distinctions*). And Flemons goes on: "A desired 'good' is isolated and pursued as if it were an independent entity, as if there were no limit to it, no threshold beyond which it stops being good, and no recognition that there is *always* another side to the coin,"[11] which of course relates to what I discussed in the previous chapter: conscious purpose is a matter of adaptation gone wrong, or going wrong.

In order to survive in the longer term, as individuals and as a species, it is crucial that we account for not only either-or relationships (in the digital or "structure" realm), but also for both-and relations (in the analog or "flux" domain).[12] We can see here clearly our first two dimensions—structure versus flux, conscious purpose versus ecological awareness—interlocking.

Then again, the practically inclined person will inevitably ask: yes, but what can we *do*? And then the circle starts all over. We seem to arrive at a stalemate of sorts. A similar paradoxicality can be found in Heidegger's tool analysis. In a sense, if we attempt to grasp relational totality—the relational whole—we must unavoidably revert to abstract rational thinking. The use of the words "relational totality" alone already entails some squeezing of it into a mold of digital either-or categories. Trying to imagine or represent readiness-to-hand can only be done in terms of presence-at-hand, the "objectifying" mode. We will see in chapter 6 how Graham Harman finds an innovative way out of this puzzle—and it is no coincidence that we will be able to trace some congruities between Harman's and Bateson's views. Bateson's doubt as to the efficacy of conscious purpose to remediate itself is also reminiscent of Evgeny Morozov's analysis of "solutionism," the discourse of casting all ills in terms of problems and solutions;[13] more on this below. In any case, we seem to be left empty-handed when it comes to finding practical avenues for change. Wider system awareness appears all but inaccessible. Are technology and its guiding "A-to-B principle" then for Bateson, as for Heidegger, ultimately an all-encompassing fate that we just have to undergo? This would suggest there is hidden within Bateson's work not only the suggestion of an instrumental notion of technology (see chapter 4), but also a determinist or essentialist concept. And "our times" even intensify the challenge. In the words of Mary Catherine Bateson, modern industry makes for a system that has the tendency to "go into runaway," as it maintains "some variables at excessive levels"[14]—an observation which must have already been relevant in 1984 when she wrote these words, but that now sounds more acute than ever, with the climate crisis at its most urgent point, while still persistent calls are being made for seemingly eternal economic growth.

So does it stop here? Fortunately, no, this is not the end of the story. Here too, we can bridge gaps. What conscious purpose is, or what we think it is, actually *emerges out of* a system of interrelating relational elements: the broader network of wider purpose. But we need a detour to arrive there, not a straight A-to-B line.

A Way Out of the Paradox?

To reiterate: whereas Bateson's definition of technology that I outlined in the last chapter had the appearance of being instrumentalist—though we immediately found grounds for nuancing that impression—the above account seems to place Bateson straight in the camp of the likes of Heidegger and Ellul, who (seem to) see little or no escape from the technological imperative. The contrast could not be bigger with the famous claim attributed to Bateson's one-time wife Margaret Mead: "Never doubt that a small group of thoughtful, committed citizens can change the world. Indeed, it is the only thing that ever has."[15] Compare with the Bateson phrase I already quoted: "We do not live in the sort of universe in which simple lineal control is possible. Life is not like that."[16] Bateson's resolution often seems to have been to plainly question the sense of deliberate, conscious action as such. His self-description in the account by Rollo May as a cork drifting on a stream, which I referred to in chapter 1, surely is consistent with that disposition.

On the one hand, his attitude should be appreciated as the result of logically thinking through the consequences of his thoughts about conscious purpose. On the other hand, ironically enough, the attitude has *itself* consequences, first and foremost of a political nature—as Berman helps to point out. He argues that Bateson's ideas potentially lead to a conservative stance: if it is assumed that any directed action is essentially in vain, the status quo becomes, by such a premise, easily legitimated. One could voice similar concerns about concepts in cybernetics more generally. For instance, Berman contends, with René Dubos, that the notion of homeostasis—the characteristic of a system of keeping its internal state in a relatively stable condition over time—leads, taken to its logical conclusion, to the proposition that "whatever is, is right" (Dubos instead calls for "homeokinesis").[17]

What is more, not merely the status quo lurks. Bateson in fact risks winding up altogether on the wrong side of the dividing line, aiding eventually the theoretical enemy. For if there is nothing we can *do*, then conscious purpose—and its dominant discourse—might just as well strengthen its footing and in the end become inevitable. The already mentioned critique developed by Morozov of solutionism can clarify the point. Morozov aims his arrows at what he perceives to be defenders of technological inevitability, such as Clay Shirky ("here comes everybody")[18] and Kevin Kelly ("what technology wants").[19] "For Shirky, things just happen—remember, it's a revolution, so all resistance is futile!—and as long as the people seem to be in charge, it all must be a good thing."[20] Against Kelly, Morozov rebuts that instead of asserting that technology "speaks," we should ask, "Who is talking?" and often we will find the answer to be: companies.[21] According to Morozov, the discourse of solutionism—and thus, we could say, of conscious purpose—tends to reinforce itself until it becomes a seemingly irrefutable truth. In light of this, clearly Bateson, for all his skepticism in relation to notions of control, gravely underestimated ideological mechanisms. (I will return to this issue in chapter 6 when discussing Bateson's relation to current political streams in philosophy of technology.)

Nevertheless, Berman's critique of Bateson seems to have largely ignored the activist, idealistic undertone lingering throughout especially his later work. Indeed, there appears to be some light at the end of the tunnel—which rebukes the suspicion that Bateson would be a pure determinist or essentialist (as he didn't turn out to be a pure instrumentalist either). Mary Catherine Bateson describes how her father at a particular moment made a turnaround of sorts, eventually seeing *some* way out of the problem of conscious purpose. In 1968, Bateson organized the "Conference on the Effects of Conscious Purpose on Human Adaptation," which—already in those early stages of awakening ecological consciousness—meant to investigate the interrelation of environmental issues and human action from the perspective of cybernetics. Mary Catherine was present to note down the proceedings, in 1972 published as *Our Own Metaphor*.[22] At first, she relates, her father took a resigned, pessimistic stance, but as the conference went on, something happened: "We spoke increasingly in terms of premises, false premises about the natural world and the nature of action. This, for Gregory, was the beginning of optimism, for it located the problem not in a necessary structure of consciousness but in learned assumptions and patterns of thought."[23] With regard to our discussion in the previous chapter on conscious purpose either belonging to the human biological setup or being a culturally learned trait, this passage seems to attest to a growing preference over the course of the conference for the latter explanation. If conscious purpose is not due to "a necessary structure of consciousness" but the result of "learned assumptions and patterns of thought," then we should be able to find ways of mitigating it by *unlearning those patterns*. Mary Catherine adds that this insight was also the prime reason for the "didactic quality" of *Mind and Nature*, published in 1979.

What is more, Bateson started to realize that the conference process itself was relevant. Although the term "solution" is obviously somewhat miscast here, Mary Catherine goes on: "He came to believe that the conference itself had come to embody a process that might supply a solution, by weaving together the conscious and unconscious thought of many minds in a single system."[24] Simplistically put: if a group of people can come together and throughout a dialogical interchange move the beacons of conceptual understanding, why should a similar thing not be able to happen on a wider societal scale? "Here we were at that point where the process in the microcosm mirrors the larger process and where intimately shared experiences are a source of broader knowledge."[25] Mary Catherine Bateson further elaborates and elucidates this in the preface to *Our Own Metaphor*, stressing the power of conversation, in a globe-encompassing sense: "The development of ways of functioning that will not destroy the viability of our planetary home will depend on conversation, understood in its widest sense: individuals in interaction manipulating words and tools, the symbols of economic exchange, political power, and passionate belief."[26] In a manner reminiscent of Richard Rorty's adoption of Michael Oakeshott's phrase "the conversation of mankind,"[27] she envisions and hopes for a "worldwide conversation."[28] And notice the remark about "political power" in the above quote—which at least attests to awareness of political structures on Bateson's eldest daughter's part. The kind of negotiation that she envisages would still imply people "manipulating words and tools." There is no suggestion of frictionlessness here. There is no such thing as a utopian state of absolute mutual understanding. The conversation, if it ever takes form, will be hard-won. To begin with, one cannot consciously strive toward it, expecting to arrive at it without deviations.

But there do exist ways. If anything can counteract or at least complement conscious purpose, it should be nonconscious nonpurpose. This sounds sufficiently trite, but as Peter Harries-Jones remarks: "Bateson insists that all conscious-purposive action should be balanced by other forms of understanding which are *not* purposeful and *not* linked to clearly defined intentions or prescriptions."[29] And he adds, in line with what has been said about secondary and primary process, that Bateson's remedy for conscious purpose converges with Freud's "royal road to the unconscious."[30] Indeed, luckily, there remain—just as with Heidegger's *Zuhandenheit-Vorhandenheit* dichotomy—gateways between the conscious and unconscious realms: a "'semipermeable' linkage between consciousness and the remainder of the total mind."[31]

Truth be told, there is still (as always!) ambivalence. Notwithstanding the "major breakthrough" Bateson thought the 1968 conference was,[32] looking back on the process of writing *Our Own Metaphor*, Mary Catherine says she became at times "terribly discouraged, feeling that Gregory's arguments must lead inevitably to nonaction," although at the same time she felt "encouraged by his emerging commitment to communicating them."[33] The same unease shows in her 1991 afterword to the book, where she speaks of "the . . . fundamental question . . . whether Gregory was right in his rejection of action to bring about change. I believe he was not,"[34] although she agreed with his overall diagnosis about what I have called the paradox of conscious purpose. Through the conference the hope had emerged that by handing people the right epistemological lenses, awareness could be taught. The beauty of and awe for nature's diversity and resilience could form the starting point. In the end, Bateson's main activist drive may have been at base, for lack of a better word, *aesthetic*. Mary Catherine: "Alienated, he was not uncommitted, for his rejection of the surface phenomena of social life was balanced by an allegiance to underlying patterns that he loved as much for their elegance as for their role in sustaining viability."[35]

That said, how do we find the gateways between conscious purpose and the wider mind, again without trying to do so *deliberately*, and—still a lingering and nagging worry—getting stuck once more in conscious purpose's paradoxicality? Bateson deals with the issue, as already suggested, via a detour. At this point his theory of learning steps in. "Learning" in the Batesonian sense is all about paradox and dealing with paradox. In essence, it is concerned with the recognition of "logical types," and takes place across several hierarchically ordered levels of abstraction. Let's go into these notions more deeply in the following section.

Learning and Levels of Abstraction

We saw how according to Bateson cognitive functioning—mind—has to do with discerning differences. Phrased in a different vocabulary: the very act of cognition is abstraction. With an already quoted phrase, "Our entire mental life is one degree more abstract than the physical world around us."[36] From a flux of events, the organism abstracts pertinent information, that is, differences that makes a difference. So all living beings engage with levels of abstraction, if only at the basic level of abstracting elementary information from their surroundings. The distinction that one may make between humans and other living beings in this regard is thus merely proportional. Humans are just capable of abstractions somewhat higher in order than other animals—as also Korzybski observed.

Like Korzybski, Bateson sees abstracting as something that humans can do in either beneficial or harmful ways. However, Bateson goes further in really deploying the notion of levels of abstraction as an across-the-board concept that can help us make sense of the wider biological world. Bateson in the first instance borrows the term "logical types" from mathematics and logic, specifically the work of Russell and Whitehead, to then apply it to matters of (social) interaction and epistemology. Bertrand Russell famously argued that a class or set is one logical type higher than its members; for example, the class "teachers" is one level of abstraction higher than its individual members, namely, all the individual teachers. Or as Korzybski would later put it, the map is not the territory—the name is not the thing named. If we confuse the two levels, a logical conflict, an error in logical typing ensues. In working out the principles of family therapy, Bateson and his colleagues saw errors in logical typing as elemental in creating double binds and pathologies. Schizophrenia in his view is the result of "maltreatments of logical typing, which we [call] *double binds*."[37]

Escaping the tangles of errors in logical typing requires what Bateson calls *learning*.[38] Learning proceeds along successive levels of abstraction. A higher level can be accessed when one acquires a consciousness of the mechanisms at work on the lower level. Learning thus can, in line with a general tendency in cybernetics,[39] be defined very broadly as the turn toward self-reflexiveness by organisms: the more self-reflexivity, the higher the logical type. Specifically, Bateson distinguishes between four types of learning. "Zero learning" is the lowest form: it is about "the simple receipt of information from an external event, in such a way that a similar event at a later (and appropriate) time will convey the same information."[40] He gives the example of "learning" "from the factory whistle that it is twelve o' clock."[41] Learning I, subsequently, can be described as any change in zero learning. Put otherwise: it comes about whenever an organism responds differently to the same stimulus at a later than at an earlier time. Learning I to that extent presupposes "repeatable context." The Pavlov experiments on classical conditioning are perhaps the most famous example.

Learning II, in turn, which Bateson also calls "deutero-learning" or "learning to learn,"[42] can only occur when there is a change in the procedure of Learning I, that is, the learning itself. It comes down to an awareness of the conditions of Learning I: the expectation of a particular context. In the case of Pavlov's dog: when the dog would at a later moment act upon the situation as though it were an experimental (laboratory) context, we would be able to say that Learning II has taken place. Interestingly, Bateson points out that this form of learning stands at the origins of what in everyday experience is perceived to be a person's "character": introvert, extravert, active, passive, optimistic, and so on. Character in this sense is the set of acquired habits repeated in different contextual settings. Bateson links this up with the psychoanalytic concept of transference, which may in itself be less relevant for my purposes here, but the following quote is interesting if only for its wording and general idea:

> This viewing [i.e., of the analyst as a father by one who is analyzed] is called the *transference* and is a general phenomenon in human relations. It is a universal characteristic of all interaction between persons because, after all, the shape of what happened between you and me yesterday carries over to shape how we respond to each other today. And shaping is, in principle, a *transference* from past learning.[43]

One may or may not adhere to the psychoanalytic vocabulary, but the way in which this phenomenon of *shaping* is described here is pertinent to the general characteristic of Learning II, in that it concerns an engraving, an anchoring of certain habits, behaviors, and modes of thinking across different contexts. In this sense, Learning II has an important survival aspect: it makes sure we don't need to adapt ourselves to each new situation; we just "stay the same." But at the same time it constitutes a handicap—when repetition becomes blind, so to speak, as is demonstrated strikingly in transference. Bateson stresses that Learning II easily becomes self-validating and is therefore extremely difficult to eradicate. At this level specifically, double binds arise. They occur when a person is faced with contradictory premises, "contraries," thus *paradoxes*, that cannot be resolved at the level of Learning II itself. Here, finally, Learning III may step in, which is nevertheless "likely to be difficult and rare even in human beings."[44] It is all about the "contexts of contexts"[45]—in other words, about networks, the larger system—and if anywhere is to be found mostly in the domains of psychotherapy and religion.

Mainly or perhaps *only* Learning III, then, can offer us gateways into systemic wisdom. Bateson points out how it's difficult and rare, but it must also be emphasized how this kind of learning, as exactly an opening up toward wider relationship and networks is anything but noncommittal. Given that it entails a reshuffling of Learning II, which helps to establish our identity, it requires nothing less than an overhaul of who we are. As Berman comments, Learning III implies "a redefinition of personality."[46] Becoming, beyond the forces of monolithic conscious purpose, sensitive to the dynamics of *"multipurposed* circuits"[47] or to different, "participatory" forms of purpose (with Berman), requires a certain leap of faith: a willingness to ponder the very frameworks that steer our thinking. Bateson: "To the degree that a man achieves Learning III, and learns to perceive and act in terms of the contexts of contexts, his 'self' will take on a sort of irrelevance. The concept of 'self' will no longer function as a nodal argument in the punctuation of experience."[48] Ultimately, what we move toward in Learning III is getting a taste of those "Eternal Verities" Bateson is after—the formal characteristics that all life and all cognition share, the "*wider knowing* which is the glue holding together the starfishes and sea anemones and redwood forests and human committees."[49]

To sum up, climbing the ladder of learning presupposes resolving the contradictions or paradoxes appearing at every level by (re)framing them from a higher level of abstraction. And as indicated before, this is a so-called stochastic process, based on trial and error, in which an organism tries out various strategies, through time, in a continuous exchange with its environment: "If now we accept the overall notion that all learning . . . is in some degree stochastic (i.e., contains components of 'trial and error'), it follows that an ordering of the processes of learning can be built upon an hierarchic classification of the types of error which are to be corrected in the various learning processes."[50] One is confronted with paradoxes at a certain level, and these can only be resolved by proceeding to a next level. But the other side of the coin is also that we need conflicts and paradoxes *to make improvement possible.* Bateson puts it in the words of one his favorite writers: "As William Blake noted, long ago, 'Without Contraries is no progression.'"[51]

Almost sideways, moreover, Bateson suggests that there also exists a form of learning that goes beyond the hierarchical ladder: learning about the *relation between* the levels. Art, humor, and play partake in such a way of learning—more on these shortly. At this point specifically, Bateson and Korzybski again part ways, as Corey Anton helps to point out.[52] On the one hand, learning for Bateson indeed implies the resolution of tension arising at one level of abstraction by passing on to a higher one, like for Korzybski progress is made by becoming aware of the process of abstracting. But on the other hand, over and against Korzybski, who seeks to achieve rigor in neatly distinguishing between levels of abstraction, Bateson argues that we can never attain such logical accuracy or purity.[53] What is more, confusion of logical types, namely, paradox, lies at the heart of fundamental elements of human communication, such as metaphor and stories. In that sense, sometimes—as previously pointed out—the map *is* the territory. Put otherwise: whereas Korzybski appears to reason from an absolute hierarchy ranging from the concrete to the abstract, Bateson seems to deploy a relative frame of thought, in which the abstract becomes each time a "concrete" again in relation to a new, higher abstraction, with some perception of difference in between. And so these concrete-abstract dynamics in Bateson become fluid structures defining each given context, the succession of which one can proceed through stepwise—if able.

At the same time, though, Bateson came to see confusion of logical types as an unmistakable source of disaster. Of course, this observation is not unconnected to considerations of conscious purpose; on the contrary. Conscious purpose stands for a one-sided engagement, an attitude limited to a particular logical type, implying disregard for higher levels. This longer quote sums it up one more time:

> The innovating [i.e., conscious-purpose-driven] species or individual comes to act as if it is no longer partially dependent on neighboring species and individuals. . . . By a process of addiction, the innovation becomes hooked into the business of trying to hold constant some rate of change. The social addiction to armaments races is not fundamentally different from individual addiction to drugs. . . . In sum, each of these disasters will be found to contain an error in logical typing. In spite of immediate gain at one logical level . . . benefit becomes calamity in some other, larger and longer, context.[54]

Purpose at base is situated at a different level of abstraction than the wider system. "The premises of 'purpose' are simply not of the same logical type as the material facts of life, and therefore cannot easily be contradicted by them."[55] Which means: without taking into account wider contexts or other levels, purpose (and thinking in terms of purpose) is simply reinforced. In short, there is a fundamental connection between purpose, levels of abstraction, and paradox.

Thus, how does one counterbalance the excesses brought on by conscious purpose? In other words, how does one remediate the paradox of conscious purpose? Since runaway processes in the context of human-societal organization are not (all) laws carved in stone, but rather acquired bad habits rooting in false premises and epistemological distortions, we can, by way of learning as the—nevertheless painstaking—procedure of passing through consecutive levels of abstraction, at least to a certain extent and in theory begin to correct ills that have emerged throughout years, decades, and ages. Such is for instance demonstrated by measures taken since the end of the 1980s to reduce the emission of ozone-depleting substances such as CFCs (chlorofluorocarbons), having led to an initial recovery of the ozone layer. However, the excruciating arduousness of the learning process is at the same time painfully illustrated by the persistent difficulty and relative inability of a long string of climate summits, pacts, and other initiatives in the recent past to substantially weigh on climate policy of world powers (this seems to be slowly changing now, as for instance China is catching up with climate warming measures; but unfortunately on the other hand the United States under the impulse of Trump has in 2017 withdrawn from the Paris climate accords).

Balanced ecological change is hard, and the paradoxicality of conscious purpose stays difficult to bypass. Paradox often remains, because levels of abstraction keep being confused—or phrased differently, dichotomies keep being interpreted "dichotomously," when it is expected that one of both sides should be preferred over the other. In order to "solve" the paradox, some breakthrough has to be forced by engaging with the relation between levels, or between the poles of the dichotomy.

Navigating Levels

We have been mapping the way out of conscious purpose. The way out, we could say, runs *between levels*. Here is our third dimension: the ever-ongoing back-and-forth between levels, contexts being superseded, or situated within a larger context. Bateson saw a couple of important "remedies" or "correctives" for conscious purpose: humility, dreams, art, religion, love.[56] All go against the grain of "lineal control." "Mere purposive rationality unaided by such phenomena as art, religion, dream, and the like, is necessarily pathogenic and destructive of life," Bateson observes.[57] Perhaps *the* preeminent domain where the "in-between" can be spotted and is practiced—though probably not always or often not consciously—is *art*.

Art can disrupt and question conscious purpose. Art's task consists in "correcting a too purposive view of life and making the view more systemic."[58] Obviously, this is much in line with a common interpretation of art, as that which by definition escapes all usefulness, in contrast to design. This sharp distinction between art and design is of course constantly being questioned in art history; take as the most blatant examples art nouveau and art deco. But still, there is some attraction to the purely structural decision of referring the artistic-as-such to the realm of the useless. Jan van Boeckel, in his dissertation on art education that also employs Bateson's insights on matters of art, illustrates the point by citing Arnold Schönberg, who states that "nothing done for a purpose could be art."[59] Which does not mean that design cannot be "artful"; it means that the design part must be conceptually separated from the art part. The former serves a purpose. The latter precisely *does not* serve a purpose. The church building could perfectly serve its functional purpose of offering a gathering place for the faithful without being lavishly adorned by spires, arches, and gargoyles. A purpose, one could state, of those adornments would be to instill humbleness into the parishioners. But if so, paradoxically, they would do so precisely because of their status of being situated beyond all use.

As suggested, art has the ability to thwart conscious purpose precisely because it deals with across-context, across-level relationships. It shares with play, humor, poetry, but also with schizophrenic symptoms the characteristic of being "transcontextual." Schizophrenics, although they may suffer from the condition, have at least this in common with the artist or the person inspired by art: for them, a thing is never just a thing. There is always a "more":[60] "It seems that both those whose life is enriched by transcontextual gifts and those who are impoverished by transcontextual confusions [i.e., schizophrenics] are alike in one respect: for them there is always or often a 'double take.' A falling leaf, the greeting of a friend, or a 'primrose by the river's brim' is not 'just that and nothing more.'"[61] According to Bateson, this "never just *this*" dynamic, if we may call it that, works on the basis of the principle of *abduction*. Abduction, in Bateson's sense, involves describing one thing or event and then looking around—possibly in other contexts and in other places—to see if we can find analogous instances falling under the rule that we worked out for our initial description.[62] This principle in fact surpasses the mere realms of art and of psychopathology; it suffuses the whole of nature, that is, of mind. "All thought would be totally impossible in a universe in which abduction was not expectable."[63] Bateson gives the example of anatomy: "the anatomy and physiology of the body can be considered as one vast abductive system with its own coherence within itself at any given time."[64] Abduction also lies at the basis of metaphor. In metaphor, something is *like* something else, but not in a linear, literal way: in an abductive way. And to that extent, Mary Catherine Bateson comments, "All organisms—not just art critics and philosophers—rely on aesthetics all the time."[65] Nevertheless, in the participation paradigm, to put it in Berman's terms, the fundamentally aesthetic character of the world was still more appreciated. We who are steeped in "nonparticipation" have all but forgotten that insight.

Now, lest one might think that in order to get in touch with those fundamental aesthetics, one needs to completely revert to the *unconscious*—correcting for *conscious* purpose—Bateson forewarns: art is not about letting the unconscious burst out; it has to do first and foremost with the "relation *between* the levels of mental process."[66] And he refers to the counterculture of the 1960s and its use of drugs in order to unleash the unconscious: "What is required is not simply a relaxation of consciousness to let the unconscious material gush out. To do this is merely to exchange one partial view of the self for the other partial view. I suspect that what is needed is the synthesis of the two views and this is more difficult."[67] Bateson tried LSD himself and was disappointed by the experience. Reflecting on it and on how psychedelics are supposed to put one in touch with "pure dream," he observes: "It seemed to me that pure dream was, like pure purpose, rather trivial."[68] We have to be very careful in not falling back to one side of the dichotomy. True aesthetics are not an issue of pure primary process, of pure flux: they are about recognizing "another mind within our own external mind."[69] Always the bridging, of the two-in-one.

So where does technology stand? Technology by nature represents the realm of the inherently useful, or at least that is how we would define it at face value—in the narrow mode. In that view, art, play, religion, et cetera on one hand and technology on the other hand turn out to be two completely different phenomena, because of their fundamentally different relation to purpose. Indeed, technology can be said to be inherently about taking a "slice" out of the world, and focusing on, laboring on, or processing that slice without any further regard for the "rest." Lance Strate suggests as much when he discusses the process of abstracting and technology, showing the intrinsic link between the two. Abstracting, as for instance in symbolic communication, heightens our efficiency. When we can use the abstract word and concept "table," for example, instead of having to literally point our finger every time to a specific table, our powers for manipulating the world and for communicating about the world (as such a matter of control and of power) are tremendously increased. We can now move about and easily talk about tables without having to drag a table around. It might be a trivial illustration, but of course looked at in general and on a larger scale, this process has far-reaching consequences:

> The process of abstracting, as it encompasses the processes of perception and symbolic communication, allows us to do more with less, and therefore represents enormous efficiencies, especially for organisms with complex nervous systems such as ourselves. And efficiency, as Jacques Ellul . . . has shown, is the basis of the technological imperative. Therefore abstracting, I would argue, is fundamentally associated with technological activity.[70]

We humans and, moreover, other organisms *need* abstracting, Strate goes on to point out. If we wouldn't abstract, the totality of our environment would submerge us.[71] And then he adds, referencing Maturana and Varela, that "all systems must maintain boundaries with their environments in order to establish and maintain their integrity of systems."[72] In order to protect ourselves as a unity, we need to draw borders around ourselves. Technology is an extension thereof, with Strate citing a striking phrase by Max Frisch: "Technology is the art of never having to experience the world."[73]

But we have to catch the nuance in this reflection. One might easily assume on the basis of this that technology is then still about "one side," the narrow-abstracting mode. However, the connection made here to the autopoiesis concept is no coincidence. Also, our analysis in this chapter of the third dimension should steer us in the right direction: technology is not just about abstracting and about heightening efficiency, it is also essentially concerned with the wider-purpose dynamics of a system, with the way that system relates to its context, how it navigates different contexts, and how it, by way of exploring relationships between levels of abstraction, is able to pierce through to wider contexts, the contexts of contexts—not only as the wider mode *an sich*, but as the interrelation, the interplay between the narrow and wider modes. In the following chapters, now that I have put the Batesonian framework in place, I will further elaborate and deepen these ideas.

PART III

THE ART OF LIVING WITH TECHNOLOGY

CHAPTER 6

Toward Batesonian Philosophy of Technology

We set out to find *technology*—having observed it to be gone. Or at least we see it developing toward inexistence, that is, merging more and more with "us" human beings, and therefore harder and harder to distinguish as something we can clearly point to. It becomes *invisible*, to begin with. Medical technologies have for a longer time kept our bodies functioning, but they're doing so now increasingly in an invasive and immersive way. How should we be able to tell laboratory-grown organ tissue from our own? It *is* our own. Machine learning algorithms interact with us without us being aware of their steering influence. "Smart" devices and apps for now seem relatively isolable and recognizable, but imagine a near-future world in which they will become ubiquitous to such an extent that they simply form our living environment. Our self-driving car knows when it's time to leave for our appointment; in fact, that appointment has been automatically generated based on our e-mail account (a functionality already even quite common these days). How would we be able to distinguish our own cognition and decision-making processes from those of the algorithms? They *are* ours. But how then to decide—to return to one of the central questions I outlined in the beginning—upon *how far* we want to take this merging?

Our common-sense understanding of technology doesn't help. We are used to regarding technology as a thing with which we achieve goals and nothing more: the instrumentalist perspective. At the same time in our cultural imagination we try to process and digest nightmares of determinist doom; look at all the stories about technological calamity. This schizophrenic condition leaves us unable to account for technology in its existential, psychological, political, economic, social, societal effects—all the more now given the developments toward invisibility, seamlessness, and fusion. Our modern, Cartesian legacy is still impairing us. It instills in us the idea that we are autonomous subjects positioned over and against free-standing objects, which we are able to control and manipulate at will. But collectively-unconsciously we seem to "feel" that something is off, that this image is not correct: technology *does* things, is not just a self-enclosed thing. It has effects that fan out into broad networks of actors, components, processes, and factors, and that we *cannot* control. This is a cognitive conflict that we can't seem to settle. Enter the nightmarish visions of all-encompassing disaster, to be counteracted with whatever remainder of "humanity" we have at our disposal.

Philosophy of technology, as said, has done much to steer right in between those two epistemological extremes. In working out their version of the central dichotomy, several approaches have attempted to marry the instrumentalist to the determinist-essentialist standpoint. Technology does have a *systemic* structure, these approaches find, namely, of impacts piercing through to many domains. To that extent the determinist stance holds true. But we can also to a certain degree control, modify, and regulate technology. To this extent it has a *thing* characteristic and the instrumentalist explanation makes sense: technology is something we can manipulate and mold to our wishes. However, the two modes do not appear to us simultaneously: we perceive technology first as efficiency-oriented, neutral means-to-an-end, and only in a second instance as network structure. And the two modes appear incompatible to a degree. We cannot be completely in both at once. We cannot just use technology unhindered, while being fully aware of its wider contextualization—or vice versa. Use is always sidestepping the full consideration of side effects, and comprehensively accounting for those side effects hampers fluent use. Implicit in this view is the idea that technology is, from an everyday use perspective, by definition at all times *more*. In particular, it is *more* than merely the instantiation of efficiency and purposive rationality. To really understand technology, we need to transcend that notion of technology-as-efficient-tool: get from the *narrow* mode into the *wider* mode.

Yet there are problems with this view, that is, with the central dichotomy. In it, the notion of efficiency gets relegated to one side, as something to be surpassed as quickly as possible, given that it marks the deficient, incomplete view. In that way the notion is lost for useful application as an explanatory principle in a wider sense, together with the concept of purpose that, partially in line with common-sense discourse, is understood exclusively within the bounds of instrumental rationality. Moreover, there is the question about how the two modes, narrow and wide, interact. How do we get from one to the other? Here we encounter unresolved issues in relation to the Heideggerian notion of breakdown, as adapted among others in postphenomenological research. Seeing technology as thing, as reified object standing before us, is a sign of the so-called metaphysical inclination. The relational constellation that technologies *really* are, on the other hand, *that* is what we should start accounting for.

But this engenders paradox. Contemporary philosophy of technology has been obliged, as a philosophical (sub)discipline, to try and define its subject matter: technology. Yet the field in doing so has actually created rather the reverse: an anti-definition—what technology "is not." It does this through showing how technology's boundaries are impossible to draw: there is always something else into which technology flows over: practices, use histories, economic factors, political interests, and so on. We saw how these theoretical tactics are strangely concomitant with real-world technological developments. Each day brings new technologies that further disrupt the presumed boundaries of our human being: neuro-implants, nanobots, big data, Internet of Things, and so on. We are becoming wrapped up in pervasive networks—"mediating (infra)structures," as Heather Wiltse terms them[1]—of algorithms, code, sensors, apps, bots, wearables, cloud services—of which we can hardly grasp what they do. With technology "disappearing" like this, we are left helpless with our common-sense instrumentalist notion of technology-as-a-thing. But unfortunately, the solution offered by the central dichotomy doesn't wholly cut it either. For, the presuppositions underlying it, namely, that technology at its most fundamental level is a relational affair and that—almost as a logical consequence—we cannot draw a dividing line on the ontological plane between technology and the human being, leaves us powerless in the face of the age-old question: to which technological innovation do we say "yes," to which one do we say "no"? If we are *ontologically one* with technology, what sense is there in contesting *ontic* fusions? On which basis should we act out that contestation, have the discussion, make our decisions?

To the extent that contemporary philosophy of technology clings to instrumentalism and essentialism in their pure forms, however intermingling them in the "hybrid" construction of the central dichotomy, as long as we keep seeing these modes as separate epistemological states between which we simplistically *switch*, we run the risk of falling back into either one of the two categories. We need another definition, another concept of technology, based in another reading of the central dichotomy. That reading I have been preparing and building up in the previous chapters. In the current chapter and the following one, we will come full circle and synthesize the gained insights. Here, I first pick up again the notion of wider purpose. Then, I return in more detail to the convergences between Bateson and contemporary philosophers of technology, by way of some comparative close reading of their respective works. Specifically, I treat again of the three main perspectives I outlined in chapter 2, to then link these up to each other in the light of the emerging framework. The investigation of how the two sides of the central dichotomy particularly interact will thereafter become our focus, and I will reach out to the work of "object-oriented" philosopher Graham Harman as a catalyst to help instigate the "reaction" between Bateson and philosophy of technology. Finally, I anticipate the next (and concluding) chapter in a closing section that makes tangible the typical Batesonian lens that we would need to apply in order to make his approach stick. For, once again, this is essential: in order to develop our new definition of technology, we need to not so much look for the "real," palpable counterpart of a certain concept in the world, but rather adjust the epistemological glasses on our nose.

Purpose in the Wider Mode

Philosophy of technology, we've seen, banished *purpose* to one side of the central dichotomy. In that capacity it takes purpose for granted. How so? Technology at base level, and commonsensically, is all about purpose: we set out to achieve goal *x* by way of means *y*. What could be more obvious? In a painstaking essay, Robert E. McGinn outlines the characteristics of the "activity-form" that technology is.[2] One of them concerns *purpose*. The main purpose that technology serves according to McGinn is "expansion of the realm of the humanly possible": technology heightens our power, strengthens our ability "to realize goals in the face of obstacles to doing so."[3]

Such definition converges with the technology-as-extension notion in the style of McLuhan: technology is an extension of the human body, of human senses or capacities. McGinn specifies there to be at least six different "modes" through which the "expansion" comes about: (1) "direct extension" (of an already existing human capacity), (2) "qualitative innovation" (offering something completely new in relation to existing capacities, e.g., airplanes), (3) "risk reduction or elimination" (allowing to do something that was possible before, but now with lesser risk, e.g., birth-control pills), (4) "improvement of performance" (doing something more easily or efficiently than before, e.g., chainsaws), (5) "substitution" (where technology takes over a task, in that way making room for the execution of another task, e.g., reading a book while one's lawn is being mowed by a robot lawnmower), and (6) "increasing the means for expression of the inner life" (enabling the aesthetic expression of beliefs, emotions, and so on).[4]

These characteristics undoubtedly form a good description of what technology can be. Technology does help us do these things. Nevertheless, as philosophy of technology has abundantly shown, if we would stick solely to this description, we wouldn't be able to explain technology sufficiently—it would be merely an instrumentalist account. Even though this definition paints a richer, more varied picture of "efficiency," all these aspects still concern merely the technical value of doing things—that is, attaining goals—better, more cheaply, more easily, or in new ways (McGinn himself does go on to discuss the value-ladenness of technology somewhat later).[5] We need to move beyond the instrumentalist, narrow side, to pierce through to the wider side of network-shaped impacts and effects. Yet, perhaps exactly because of its allergy to instrumentalism, philosophy of technology has distanced itself from truly dealing with purpose. Indeed, most clearly in the narrow mode, technology is, also in an everyday sense, defined through "purpose." We have all-purpose cleaners, for example. We ask ourselves constantly what things' purposes are, and we look for the right tools to achieve the purposes we envision. This is lost when we turn to the broader mode of network constellations. Here, the idea of purpose evaporates.[6] There cannot be a grander purpose in this network structure, we seem to say—just like according to the orthodox scientific paradigm we're not allowed anymore to allot purpose to living systems.

However, looking at the wider mode, we can find purpose just as much. There as well, purpose is ubiquitous. Putting it even stronger: *it might be even more important there than anywhere else*. We just lost the habit of thinking about the wider mode in that manner. What we need to teach ourselves is to think about technology in systems-theoretical terms of *purposive structures*. Bateson offers the platform for such an undertaking. We saw how his approach is grounded in a cultural critique, namely, of conscious purpose. Conscious purpose in Bateson's elaboration, as we've found, converges with an instrumentalist attitude, but also to a certain extent—looked at from another angle—with the determinist view. This in itself should be a sign. Conscious purpose gets enhanced by technology; technology is the materialization of conscious purpose, Bateson asserts. This seems a simple instrumentalist notion. But when he appears to suggest, on the contrary, that conscious purpose for us humans forms a kind of inescapable fate, he tends toward a determinist view.[7] Yet he also points out how we *need* technologies, even to counterbalance the very ills of conscious purpose, and then he appears to side fully with contemporary philosophers of technology who regard technology as ontologically intertwined with humanity. This all seems a bit confusing; however, it provides the key to the issue: we have to remind ourselves of our deeply ingrained habit of thinking of technology as a "something"—and turn that reflex on its head. Technology is not the main issue; *purpose* is. It is about the linear, monomaniacal pursuit of goals in the conscious-purposive disposition *versus* the awareness of wider purposes, that is, wider ecological (im)balances. And to that extent, as we just saw in the previous chapter, it is about sticking pathologically to one logical type or level of abstraction, versus taking into account multiple levels and, crucially, the relations between them. It is about single-purposiveness versus getting attuned to multipurposiveness.

Now the problem is not that we want to understand technology as "for" something—on the contrary. That is only a problem if we restrict that explanation to the narrow side alone, as instrumental rationality. We have to (also) learn to think of technology as purposive in the wider mode. But this will imply that we reshuffle our image of how we think the connection between the two modes, the two sides of the central dichotomy. *It does not mean*—this is crucial to emphasize—*a (re)importation of instrumental rationality into the wider mode.* Rather, we should envision a two-sided structure that, as a constant of existence, shapes our condition at all times, shapes the way we look at the world. The central dichotomy does not constitute a set of two modes to *choose* from, depending on how narrow or wide our view is (do we just use technology functionally-instrumentally *or* become mindful of its expansive impact?). No, the two sides or modes are always "there," and it is the form of the interaction between the two that colors and configures our situation. Which type of interaction currently reigns, depends on the level(s) of abstraction we are "on."

Indeed, technology along these lines should be conceptualized not as *technology* in the first instance, but as a matter of *purpose*—however, again not in the narrow, purposive-instrumental sense alone. As far as we humans are ontologically continuous with technology, we are beings who cannot but be technological. In this sense, technology already "disappeared" a long time ago, if it was ever there ("we have never been modern," we are "natural-born cyborgs," et cetera).[8] And on an ontic plane, it is disappearing now in even more dramatic ways—it becomes merged in tangible and sometimes frightening ways with our bodies and minds. So how to make sense of these mergers? Answer: by saving "purpose" (*and "efficiency"*) from its isolation on the narrow side, and casting it anew as a central principle to help us understand technological decisions: what do we want to *do*, *achieve* with technology, and with specific technologies? *Which goals do we want to attain*—not merely in an instrumental-rational sense, but also psychologically, economically, politically, socially? To differentiate between on the one hand conscious-purposive (instrumental-rational) perversions and on the other hand multipurposive, multilevel possibilities becomes then the ultimate "art of living" skill.

Before we further unpack these ideas in this and the following chapter, we must return to the intersections between the Batesonian conceptual universe and the contemporary philosophy of technology in detail.

Bateson and Philosophy of Technology

Generally speaking, it shouldn't come as a big surprise that Bateson's thinking lines up well with aspects of twentieth-century philosophy and contemporary philosophy of technology. As we've seen, both Bateson and philosophy of technology make it their central task to react to the Cartesian paradigm. As a result, Batesonian notions, as well as hints or variants of ideas that this book puts forward, can be seen to be either wholly apparent or lying dormant within the field.

Take the work of one of the pioneers of contemporary philosophy of technology, Albert Borgmann. Borgmann famously distinguishes between the "device paradigm" and "focal things and practices." The device paradigm is the main "lens" through which we view the world nowadays. We see technologies as merely devices, merely *means*, while our consideration of *ends* dwindles. Borgmann elaborates the notion largely on the basis of Heidegger's *Gestell*. "Enframing," as way of thinking and being, reduces all things to a resource for production and consumption, disregarding completely their place and meaningful role in a world, or, their world-disclosing character. This kind of thinking, Borgmann argues, defines our contemporary culture and worldview. In opposition to this, he places a call for an enduring engagement in focal practices and with focal things. Unlike devices, which harbor an "emptiness," focal things ask for our continuing concern and care.

Borgmann gives examples such as running and the culture of the table. These activities and practices ask that we focus on them fully to make them worthwhile. Thereby they "make" a world. "A focus gathers the relations of its context and radiates into its surroundings and informs them."[9] But this also implies a completely different relation to ends and means. Focal things and practices are not merely means; their ends define what they are.

Based in these reflections, Borgmann puts forward a societal "reform proposal." *Ends* should again be evaluated, in themselves, and in relation to the means. He is on the lookout for a "broader sense of the means-ends relation" that is clearly "in conflict with the means-ends structure, embodied in the device paradigm."[10] But he sees the trap that Bateson also spots in the context of conscious purpose. "New ends" should be introduced, but in the form of focal things and practices, not of yet again "different commodities." That would just be a continuation and extension of the device paradigm. Borgmann remarks with regard to other attempts at mitigating "the excesses of technology":[11] "Most traditional reform proposals are finally ensnared by the device paradigm and fail to challenge the rule of technology and its debilitating consequences. Hence a radical reform . . . requires *the recognition* and the restraint of the device paradigm, a recognition that is guided by a focal concern."[12] Such admonition is partly reminiscent of the difficulty and paradoxicality we observed in attempts to bypass conscious purpose: in wanting to "solve" conscious purpose, we risk reverting to the same dynamic we are trying to escape. And, indeed, in his program for reestablishing our attention to wider contexts, Borgmann proposes a way out that resembles Bateson's levels and learning approach: we need to become attuned again to how things are taken up in their environment, how they are part of a wider significance. This means, rephrased in the vocabulary of this book, attending to purpose in the wider sense. Nevertheless, when Borgmann stipulates that the goal of his reform proposal entails "*to restrict technology to the status of a means* and to introduce new ends,"[13] he seems to reverse hard-won insights (as in general also pointed out by, among others, Peter-Paul Verbeek: Borgmann to some extent represents a throwback to classic determinism).[14] This is casting technology again to one side and "meaningfulness" to the other, wider side in a sort of Habermasian maneuver (see below)—with the command to strictly guard the borderline between the two from now on: "Let them not intermingle!" We've seen, of course, that such a purification act, in Latour's terms, can only be executed in words or in theory. Technology is a matter of both sides: narrow and wide. But *purpose* is as well. Admittedly, Borgmann's intentions are good enough: he endeavors to put technology "in its place"—no longer treated as an end unto itself—and again in the service of well-chosen, well-debated ends. But as such, and certainly in the light of the technological developments that form the backdrop for my study here, that is, of technology "disappearing," this is unfortunately not completely helpful.

Another approach, closely related to this one, though sufficiently different, can be found with the French philosopher of technology Bernard Stiegler. Stiegler adopts the term *écologie de l'esprit*, or "ecology of mind," from Bateson. In the English translations of Stiegler's books, the term is then retranslated as "ecology of spirit," which cloaks its origin—but in any case Stiegler does not systematically engage with Bateson's work, and even scarcely refers to it (there is a passing reference to Bateson's theory on alcoholism in *What Makes Life Worth Living*).[15] "Ecology of spirit" in Stiegler's work has, for a part, a pejorative flavor, coupled as it gets to a "crisis" that Stiegler outlines and that relates to the current reign of digital technologies, "technologies of spirit." Insofar as these serve to "proletarianize" us, reducing us to mere consumers of information, services, entertainment, and so on, obliterating our "savoir-faire" and "savoir-vivre," they contribute to the crisis. But as always in Stiegler's work, the *pharmakon* that technology is offers poison and cure at the same time: exactly these technologies of spirit can aid in liberating, "de-proletarianizing" us.[16] In that sense the "ecology of spirit," as our living environment so to speak, is something we need to carefully watch over and care for.

To this extent Stiegler's work indeed coheres with Bateson's. This is also demonstrated by Daniel Ross in a paper titled "The Ecology of Spirit: Technics and Politics in Bernard Stiegler," that compares Stiegler's and Bateson's perspectives in the light of their account of "mind" or "spirit."[17] In both, Ross finds adversaries of the "cognitivist" view, that understands things solely "in terms of bits of information,"[18] and that is specifically criticized by Stiegler. Bateson would appear at first sight a cognitivist, given the way he works out his founding concept of "information." However, Ross suggests, he offers more. Bateson elaborates on the basis of his concept of mind and the notion of conscious purpose what Ross calls a "cybernetic phenomenology."[19] Ross, too, notes technology's role, for Bateson, in aggravating the dangers of conscious purpose. And he helps to point out that there is a "historical" aspect to this. Human individuals and societies are able to transmit their knowledge, worldviews, and habits across the generations.[20] This means that throughout time, an "ecology of bad ideas," in Bateson's words, can arise.[21] Ecologies can grow "bad," due to all the distortions taken together. "Bateson suspects that this cumulative character of the influences of perceptual selection [i.e., as engendered by conscious purpose, the short-circuiting of the loop structure to which I referred in chapter 4] somehow makes possible a systematic distortion of consciousness."[22] Ross only implicitly here suggests the analogy with Stiegler's central notion of *hypomnesis*—the exteriorization of memory and cognition in mnemotechnical devices, enabling transmission through time, but also more fundamentally constituting technology as temporal, and the human being as technological—but the parallel is clearly there.

In furthering "an ecology of good ideas," Ross indicates, Bateson develops something of a "politics." Only, he adds, "one searches in vain in Bateson's work for detail of what this might mean in practice."[23] Stiegler, by contrast, goes further in this, according to Ross. In the terms that he adopts from Gilbert Simondon, Stiegler diagnoses the current "ecology of spirit" as "disindividuating" the person: the person (as consumer) is "devalorized,"[24] regarded and treated merely as consumer; whereas ideally we should strive, with our technologies, toward "psychic and collective individuation."[25] In contrast to Bateson, Ross argues, because Stiegler accounts for this, his politics are more powerful, notwithstanding the initial affinities between their two accounts of mind/spirit:

> Like Bateson, Stiegler understands the genesis, flourishing or floundering of individual and collective mental processes in terms of a kind of ecology. Whereas Bateson speaks of an ecology of bad ideas, however, for Stiegler what matters more than the bad ideas is the quality of the ecology itself—bad ideas, or no ideas, are the *result* of a bad ecology, that is, of a system that leads not to psychic and collective individuation but to their opposite, to disindividuation. Politics, today, according to Stiegler, can consist only in the struggle against *this* tendency and its multiple consequences.[26]

As we've seen, this is not an isolated instance of the critique of Bateson's lacking political perspective. I will return again to Bateson's politics shortly and argue that notwithstanding shortcomings, his approach in my view *can* offer us unique politically relevant insights, beyond dominant categories in philosophy of technology. Let me for now conclude this discussion on Stiegler with one more brief addition to the comparative analysis between Stiegler and Bateson, more precisely with regard to purpose and technology.

In *Technics and Time, 1*, Stiegler comments on the Habermasian distinction between "purposive-rational action" and "communicative action," with the former in Habermas's perspective encroaching upon the latter. Stiegler points out that Habermas faces "the same paradox concerning technical modernity" as Heidegger:[27] technology (in Stiegler's terminology "technics") is a human power that serves humanity, but it gains an autonomy *as* it empowers us, "as a result of which it does a disservice to active humanity."[28] But Habermas and Heidegger arrive at different analyses, Stiegler goes on. "Habermas continues to analyze technics from the perspective of the category of 'means,' a category considered by Heidegger to be a metaphysical determination."[29] In Heidegger's view, indeed, we cannot linger with the instrumentalist explanation of technology as only means-to-an-end: technology is *much* more, if not "everything," at least in our times, in the sense of technology completely conditioning existence. Stiegler will go on to adapt Heidegger's reading of modern technology but also to radicalize it. Technology is not the latest occurrence of metaphysics; it is "originary." This way, de-essentializing Heidegger we could say, Stiegler sculpts his own version of the central dichotomy, continuing philosophy of technology's dominant theme of ontological continuity between the human being and technology. In this capacity, moreover, by way of its notion of "originary technicity," which, however, still constitutes and conditions human life, Stiegler's work can be usefully deployed in order to counteract too-instrumentalist accounts,[30] of the kind we indicated as potentially problematic in the introduction. Nevertheless, together with discarding Habermas's too-rigorous split between purposive-rational and communicative action (and rightly so), Stiegler also casts out a specific treatment of *purpose*—notwithstanding that his overall *cultural* analysis eventually goes a long way into the direction of what I am envisioning here. This is a difference with Borgmann: while Borgmann singularizes technology (as device paradigm) but reimports purpose, Stiegler "flattens" technology—technology *is us*—but loses the grip on purpose. Often, Stiegler seems to compensate just a bit too much on the other end of the spectrum, arriving at a sort of generality, not faulty in its abstractness, but missing as such a crucial characteristic of the central dichotomy, namely: the *interaction* between the two sides (while Borgmann, more in a classic-essentialist vein, *severs* the two sides again: the device paradigm and focal practices are two qualitatively different things, wholly incompatible with each other).

We can further conceptualize these issues through the notion of *breakdown*—I will do so in a short while. Before that, we need to revisit once more the three main approaches within philosophy of technology that I presented first in chapter 2.

Postphenomenology

I already remarked upon how Bateson's relationalism concurs with postphenomenology's relationalist outlook. In the postphenomenological tradition, mediation is seen as co-constitutive *of* entities, of subjects and objects, if these terms are even applicable. There isn't such a thing as a loose in-between that is added—theoretically or practically—to entities. Mediation *makes* its poles. We've seen how Ihde points out that equipment—readiness-to-hand, "relational totality"—is primordial. Bateson, in turn, puts in a similar vein that "the relationship comes first; it *precedes*."[31] At one point when he discusses patterns of interaction between individuals of different species—play between a dog and a gibbon, or between a dolphin and a human—he observes that although each creature remains "itself" in the process, something *more* is also created, and this he calls "practice": "There is . . . a larger entity, call it *A plus B*, and that larger entity, in play, is achieving a process for which I suggest that the correct name is *practice*."[32] Read a bit differently, this statement could just as well apply to human-technology relations as depicted by postphenomenology. Obviously, Bateson and Ihde are speaking about different things: Bateson talks about interacting animals, Ihde zooms in on the relation between humans and technologies. But their underlying conceptual infrastructure is alike: through relation, something happens, something new comes forth, rather than the reverse, that is, that preexisting things make a relation.

What shines through here also is a pragmatist motivation, in postphenomenology's case of course more deliberately sought after than in Bateson's. Knowledge in the pragmatist view arises through practice—remember, preeminently a relational affair—and Ihde translates this insight into an analysis of how technologies mediate knowledge production, most famously in his study of Galileo's telescope and on a more conceptual level in his elaboration of the notion of "instrumental realism."[33] Technological instrumentation has always had primacy over "pure" scientific abstraction, not the other way around—a state of affairs, according to Ihde, that Heidegger observes but that Husserl misses. Galileo "needed a telescope" in order to subsequently produce his theoretical abstractions. Husserl refused to acknowledge that properly. "Only through being technologically mediated is the newly produced knowledge possible," Ihde claims.[34] Technologies to that extent act as "epistemology engines": devices "that bring human knowers into intimate relations with technologies or machinic agencies through which some defined model of what is taken as knowledge is produced."[35]

Bateson does not purposely refer to pragmatist traditions, but affinity can be spotted with his overarching epistemological perspective: knowledge comes about in practical processes of doing. As we've seen, mind in Bateson's view is not the abstract cognitive set of capacities we usually imagine when we think of the term. It is organisms' central way of getting around in the world—an insight later incorporated in theories such as enactivism, embodied/situated/distributed cognition, the extended mind, and the like.[36] Notice also the kinship with the Deweyan approach in philosophy of technology elaborated by Larry Hickman. Hickman conceives of technology in terms of John Dewey's concept of "inquiry."[37] Technology in his definition is the "invention, development, and cognitive deployment of tools and other artifacts, brought to bear on raw materials and intermediate stock parts, with a view to the resolution of perceived problems."[38] Developing and deploying technologies is part of "inquiry," conceptualized by Dewey as "productive skill whose artifact is knowing."[39] And inquiry is always tied to a particular situation. In this perspective, technology makes for an across-the-board human activity, something that we *do*, but that also in the process shapes our way of looking at the world.

That doesn't mean the central dichotomy doesn't "hold" in this context. We can expect a unilinear A-to-B solution from technology (narrow mode) through this activity, *and* we can become sensitive to other, broader purposes (wider mode). In fact, postphenomenology's *multistability* concept constitutes its central way into bridging the gap.[40] Against what Ihde calls the designer's fallacy, namely, the idea that a technology merely is used as the designer foresees it—thus merely serves the purpose the designer puts into it—multistability shows how technologies are open to *multiple* purposes. Multistability would thus *in potentia* enable the crossover from the efficiency-oriented to the wider-purpose mode. As mentioned previously, postphenomenology traces down transparency-opacity ratios in several different use contexts. However, as it stands, in applications of the postphenomenological framework, investigations of the multistability of technologies have stayed limited mostly to (the in itself worthwhile effort of) searching and finding *other* purposes apart from the obvious, known, or designed purpose of a technology. They are not explorations of multipurposiveness in the way which is outlined and advocated here—that means, crucially, at different levels of abstraction (more on this to follow).

Actor-Network Theory

An analogous narrow/wider dynamic characterizes Latour's black box notion. We saw how Bateson seeks to solve the mind-matter problem. His approach means to close the chasm between mind and matter, by seeing them in one vision—the technique of double description. We need to deliberately see double: mind is the making of marks within the flux of matter. Nevertheless, our modern Cartesian upbringing is persistent and keeps plaguing us. Implicitly, automatically, we still tend to think of the two (subject and object) as separate realms, just like we, according to Latour, still want to explain phenomena as either "social" (subject) or "natural" (object), but fail to see that we have always mingled the two. We have always made hybrids. We have always made black boxes.

Bateson uses the term "black box" himself, in one of his "metalogues": playful imaginary dialogues between a father and a daughter (partly inspired by real-life conversations between him and his daughter Mary Catherine; specific to a "metalogue" is that its structure, the way in which the dialogue is held in whole, is relevant to the subject under discussion). Daughter asks: "Daddy, what's a black box?" whereupon Father answers: "A 'black box' is a conventional agreement between scientists to stop trying to explain things at a certain point. I guess it's usually a temporary agreement."[41] And he continues to explain that the term comes from engineers, who use black boxes as shorthand when for instance drawing diagrams: "Instead of drawing all the details, they put a box to stand for a whole bunch of parts and label the box with what that bunch of parts is supposed to *do*." Whereupon Daughter summarizes: "So a 'black box' is a label for what a bunch of things are supposed to do." Answer of Father: "That's right. But it's not an explanation of *how* the bunch works."[42]

No true conceptual overlap should be sought here with Latour's much richer notion, yet Bateson's description is essentially correct, notwithstanding that it is formulated for the specific context of science-making (an area that Latour of course covers as well). But most importantly, the description denotes the wider observation by Bateson about how epistemology works: by "pinning down," that is, making distinctions within the stream of flux. The image of a black box as a temporary grabbing together of a bunch of things, making them appear as "one," resonates strongly with Latour's perspective in which what are in actuality hybrid assemblages are able to pose as well-circumscribed things.[43] As we'll see later in the discussion of

Harman's object-oriented ontology, the groundwork of Latour's theory is every bit as relationalist.

Critical Theory of Technology

Latour's black box concept is then adopted by Feenberg, as we saw, who expands it toward a Critical School–inspired political philosophy. We delineated how Feenberg describes the existence of a "technological unconscious," to which values get "repressed," belonging from that moment on in material form to the accepted design of a technology. Other values by contrast may still circulate in discourse, that is, be more easily accessible to conscious debate. But the black box is much harder to deconstruct. Think again of the bicycle case examined by Pinch and Bijker. We perceive our current bicycle design as normal, and ask throughout our use of it no longer what the social values—political, economic, cultural—implicit in its design are. And through constituting black boxes, Feenberg adds, technology is able to consolidate reigning power relations, in favor of dominant groups. This represents Feenberg's narrow side on the central dichotomy. Primary instrumentalization allocates objects and subjects as elements of a technical system. They are pulled out of their life context and reduced to pawns, as a matter of speaking, in the game of technical rationality, that serves in turn to solidify power constellations. Secondary instrumentalization, conversely, enables a reintegrating movement that assigns existential meaning to the technical object and subject placed again within specific lived contexts. This "subversive rationalization," instigated by "technical micropolitics," deployed by individuals, interest groups, or organizations from below, *recontextualizes* technology—thus, bringing us to the wider side of the central dichotomy.

We find back here Bateson's dynamic of climbing ladders of abstraction, *learning* through "contextualizing" each consecutive layer, that is, placing the current level in a wider context and thereby becoming able to transcend it toward a higher level. This may not seem straightforward from Feenberg's work at face value, but it can be made tangible by a brief detour through one of his main influences: Marcuse, and specifically, *One-Dimensional Man*.[44] Marcuse in this iconic work shows quite lucidly how the counteracting of technical rationality has to do with distinguishing between "abstract" and "concrete." For it is "the concrete" that is everywhere imposed as the ultimate touchstone of societal activity—up until today, we may add, and perhaps today even stronger than ever.

Marcuse famously describes *one-dimensionalism*, the process by way of which capitalism is creating a society and a type of human being unilaterally bound to senseless work, blind consumerism, and profit-making. In his investigation, he reserves a central place for a close scrutiny of the ways of thinking about science and knowledge that are encouraged or, rather, imposed in such a constellation. One-dimensional society, in Marcuse's view, tends to be radically functionalist. Identifying things completely with their functions,[45] its physical sciences are operationalist and its social sciences behaviorist in spirit, and both are grounded in a "total empiricism."[46] By implication, such a system "militates against abstraction."[47] There is one big reason for this: abstract thinking is dangerous. It threatens to undermine one-dimensionalism, for it is through critical, abstract thought that we can enter a "two-dimensional universe of discourse" again.[48]

Nevertheless, the situation is not that simple. Something is going on with the categories. One-dimensionalism preaches concreteness and empiricism: a certain down-to-earthness, one might say (as mentioned, still very much in vogue today). But according to Marcuse this is a "false concreteness," "a concreteness isolated from the conditions which constitute its reality."[49] This specific form of concreteness serves to keep the individual in check, supportive of the system that, after all, "delivers the goods."[50] It obscures its own political underpinnings, making itself in this way immune to refusal and protest.[51] What we need in order to oppose this are a "negative" language and way of thinking, able to expose the mechanisms behind this construction of the so-called "concrete." As stated, only abstract, conceptual thought is capable of "recognition of *the factors behind the facts*."[52] Over and against "the fallacious concreteness of positivist empiricism,"[53] political philosophy should seek "to comprehend the unmutilated reality."[54]

Concrete and abstract are strangely intertwined here: "the concrete" is not really the concrete, just a function of ideology, a mirror of an imposed abstraction, that as such closes the gateways toward critical abstraction "of the good sort." Indeed, Marcuse says, although "abstractness is the very life of thought, the token of its authenticity," still "there are false and true abstractions."[55] Analytic philosophy, for one, is in Marcuse's view guilty of the former. Conversely, critical political philosophy achieves true abstraction, in that it "abstracts from the immediate concreteness in order to attain true concreteness."[56] True concreteness then concerns the way in which society is *really* constructed—according to fault lines of power consolidated through technologies' "technical code," in Feenberg's terms—not just how discourse *tells us* it is.

In this way, Feenberg's call toward subversive rationalization can be seen to converge with the Batesonian call toward wider ecological awareness, by navigating levels of abstraction. It is essentially a matter of *widening the view*. However, also precisely here, an important difference among the approaches in philosophy of technology that I have discussed, emerges.

So far we have employed our three main approaches in the service of the investigation without much going into their mutual relations. Yet there is a specific relation between them, making them naturally fit together as they map onto the central dichotomy. Actor-network theory (ANT) can be looked at as a middle piece in between postphenomenology and critical theory of technology. Postphenomenology in the work of Verbeek has coopted the actor-network perspective, in order to constitute a "mediation theory."[57] And as pointed out, critical theory of technology has been influenced by ANT in adopting the latter's concept of black-boxing.

Each of these two, postphenomenology and critical theory of technology, has its own orientation in relation to ANT and to the central dichotomy. The three approaches can be said, then, to each cover slightly different ground. Very simplistically put, one could say that the postphenomenological outlook pure and simple—much in a phenomenological tradition—looks first and foremost at the experiential standpoint, that is, of a technology user, in a specific situation. *Use* is meant here in the generic-phenomenological sense of any deployment of technology looked at from the standpoint of the "observer"; this may also include "unconscious" use, for example, of brain implants. Actor-network theory by definition goes beyond that user standpoint, to map, as comprehensively as possible, the endless networks of actants in which a user-observer is only one element. And Feenberg builds on ANT, but adds a component: an overarching theory of social justice.[58]

There is a longer-standing debate between postphenomenology and critical theory of technology that helps to illustrate the lines of force that are at play in between the perspectives. Postphenomenologists have criticized Feenberg for failing to grasp the interwovenness of the human being and technology and for still being founded upon inherently modern principles.[59] Feenberg in turn faults postphenomenology for furnishing a framework that is lacking in political and moral relevance.[60] The debate is friendly, though, with much goodwill exhibited on both sides. Also, it could be said there is a certain openness to both approaches that makes them highly suitable to adaptation or completion from the outside.[61] More recently, a group of Dutch scholars has again attempted to correct and complement the postphenomenological outlook. Mithun Bantwal Rao, Joost Jongerden, Pieter Lemmens, and Guido Ruivenkamp begin where the aforementioned debate with Feenberg left off, averring in fact that critical theory of technology and postphenomenology—though they zoom in on Verbeek rather than on Ihde—have more in common than mostly assumed, to the (pejorative) extent that both remain merely "reactive" in their recipes to resist the "power of technology," and do not question the locus of technological production itself (an ailment the authors propose to cure by adding autonomist Marxism to the mix).[62] Dominic Smith, in turn, pleads for a form of postphenomenology in which the reputation of "the transcendental"—exactly a Husserlian concept that Ihde seeks to largely discard[63]—is restored, in that manner enabling a better grasp of "constitution," that is, the conditions that constitute technology.[64] Smith points in the first instance to political constitution: an aspect, as was already clear from earlier discussions, that is said to be largely ignored by postphenomenology. Smith's argument in a way links up nicely with that of Rao et alia, as the latter localize those conditions first and foremost in the relations, forces, and means of production under capitalism.

Notwithstanding these and other efforts to bring the two camps closer to each other,[65] however, a chasm remains. This was lately again illustrated by debates held in relation to a recent attempt by Verbeek to build a political philosophy within, or on the basis of, the theory of technological mediation.[66] Verbeek sees in critical theory of technology a remnant of the Marxist tension between oppressors and oppressed, in relation to which, he claims, only "resistance" can be a viable coping strategy.[67] And it is exactly this kind of dualistic thinking we need to get rid of, he argues. Verbeek instead proposes an "ethics of accompaniment" that finds a middle road between rejection and blind acceptance.[68] Feenberg from his side is sympathetic toward Verbeek's approach, but on one hand rejects the latter's interpretation of his framework as overly dualistic,[69] and on the other hand continues to make a strong plea for the acceptance of the simple fact that (grave) power asymmetries *do exist*, a given that postphenomenology seems to want to ignore. Most recently, Feenberg has stated this with a quip—specifically referring to STS, nonetheless, but the remark could just as much apply to postphenomenology: "Any method which fails to recognize the widespread existence of deception and corruption is fatally naïve. (Volkswagen has a car for those who dismiss the critique of hidden motives as outdated.)"[70]

This remains a seemingly insurmountable stumbling block in the mutual rapprochement. Verbeek has declared in (the aforementioned) discussions that critically oriented commentators have it "easy" to complain about powers that keep us in check; they don't "do" anything. The underlying assumption is that more pragmatically inclined perspectives—as offered typically by many Dutch strands in philosophy of technology—with an affinity toward design, engineering, research and development, and so on—fields that deliver "concrete" results—are able to have much more impact in the political realm. These approaches can *really do* politics: influence societal organization by developing tangible solutions. Merely talking about power imbalances, as critical theorists are wont to do, in the worst case comes down to, with a term Verbeek has wielded in debates, "intellectual masturbation," saying: "It's nice for you but you won't make the world a better place."[71]

We can return here to Bateson's political view, or his purported lack of a political view and his supposed conservatism. In his (initial) unwillingness to act, in order to "solve" the ills of conscious purpose, concerned that the solution would only protract the problem or even make it worse, Bateson does indeed seem to hold on to a conservatist stance. Nevertheless, in the light of the discussions here, one could also turn the issue upside down. *Action* does not by definition, always and everywhere, constitute the *progressive* reflex. In fact, in neoliberal times, calls for "concrete" actions, solutions, and interventions should be treated with the greatest skepticism; remember Marcuse's critique of operationalism and functionalism. If our ideological self-understanding is cast completely in these terms, we'll have a hard time looking beyond those categories. Developing, designing, making, producing (and consuming) then seem like the most natural, most logical options. But in fact *doing*, in this capacity, can make sure that the existing situation is only maintained, confirmed, and bolstered. So looked at through a Batesonian lens in this way, *doing nothing* might actually be, rather than conservative, the *progressive* course of action, or more precisely, nonaction. In the face of the neoliberal-individualist frame of mind that eternally celebrates action and entrepreneurship, the Batesonian, Taoist-like *refusal to act* offers a refreshing incitement.[72] While typically the obsessive "act now!" solutions-driven attitude is in risk of negating side effects of all sorts, this Batesonian political stance calls on us to become aware of as many side effects as possible. One could even suggest in this perspective, and a bit provocatively, that perhaps (real) masturbation is what designers, developers, policy makers, entrepreneurs, and so on should do *more* now and then—instead of storming the sky for the nth time.

Bateson can act as a kind of glue here, in between postphenomenology and critical theory of technology. Yet the fact remains that these perspectives keep differing in the foundations of their worldview—and this carries over into the positions that they take within the central dichotomy. The central dichotomy in general traces the contours of a fundamental *visibility-invisibility* dynamic characterizing technology. Something appears about technology—its function as a means to an end—while something else hides from view—its network structure of impacts and effects; or vice versa. But among our approaches, a distinction needs to be applied that has to do with the specific *type* of (in)visibility that is at stake. As said, postphenomenology takes the user perspective, the perceptual-experiential perspective, as its starting point. Critical theory of technology focuses more on the systemic perspective of power relations and (im)balances. The type of (in)visibility relates to these foci.[73] Feenberg works out a view in which the invisibility mainly concerns the "context" of technology use. In using a technology, we stay oblivious of all political-power constellations constituting it. For postphenomenology, however, the invisibility resides mostly in the use itself: in using a technology, the relational character of that very use escapes us.

The tension between this use and context mode seems unbridgeable. It indicates a dimension, a *cut* running perpendicularly across the central dichotomy. It also lies at the base of the concern that more pragmatically oriented approaches such as postphenomenology are in risk of reverting to instrumentalism. Indeed, when one takes the situation of "use" as one's starting point—again, use meant here in a generic sense, as standpoint of an observer (who doesn't necessarily always *observes* literally; the point of being the observer can also entail that one *doesn't* observe certain things)—one is in danger of neglecting wider systemic factors; *even though one tries to map "phenomenological relationality."* One crosses the central dichotomy all the same, but on a different *level*. Through Bateson, however, we've gotten used to "extra" dimensions of *two*. In fact, it is looking through the Batesonian lens that we can easily perceive these *twos* as part of the same *one*. We start seeing things as inherently double-sided. Exactly the Batesonian perspective links up use and context in one and the same view. We *act*, chasing after our conscious purposes. We have to act, for survival (this in the end might be *the* main purpose of technology as such, on the most general plane). But survival—cultural, economic, political, psychological—also necessitates a guarding attitude with regard to the excesses of conscious purpose, as exactly these start to impede that survival. And this involves a view on wider phenomenological relationality as well as wider systemic-political (i.e., Feenbergian power asymmetry-related) relationality. *Both* belong to the "systemic wisdom" that Bateson advocates. *They are just different levels of abstraction*—mapping out onto the Batesonian dimension of logical types.

"Just" in this sentence sounds, admittedly, a bit too quick. How then do we get from one level to the other? This links up to that other problem that I described, plaguing contemporary philosophy of technology's central dichotomy: the unclarity concerning the process of getting from one mode to the other. How does the switch happen between narrow and wider mode and vice versa? Remember again how on the basis of Bateson's model, we must regard the relation between the narrow and wider mode as a difference of levels of abstraction. The digital, or *map*, is one level higher than the analog, or *territory*. But we can go on abstracting, in a sort of two-dimensional frame—with the central dichotomy in its general form as one dimension, and levels *across* that dichotomy as another one. The topological metaphor might be misleading. We are not really concerned here with dimensions as in a physical or mathematical multidimensional model. Rather, we are involved with the experiential-existential-epistemological point of view. This is what we are essentially interested in: how to cope with technology in everyday life, when technology is a disappearing thing? In this sense, our experiential viewpoint can only be *one*. We can only be "our" person. *But* we can skip locations, change places; take different viewpoints, depending on where we stand. And this is what the notion of *breakdown* really is about in this framework: changing existential perspectives—if only imaginarily. Now, all this sounds terribly abstract—too abstract—but through an unexpected detour, namely, via Graham Harman's object-oriented ontology, we can further elucidate these suggestions.

Relation, Substance, and Speculation

In chapter 2, I elaborated how, in the (post)phenomenological outlook, the "jump" from *Zuhandenheit* toward *Vorhandenheit* happens through breakdown. Heidegger's tool analysis has served as an important building block for contemporary approaches in philosophy of technology—offering us literally a way to look at *tools*, and how they either appear to us in objectified, reified form *or* are experienced within the relational structure of their "in-order-to." Indeed, both modes are purposive in their own way: the former in the narrow instrumental-purposive sense, the latter in the wider-purposive sense. We saw how Ihde finds this distinction to be crucial, and regrettably collapsed and lost in the later Heidegger. However, throughout its history of adoption, the image of the tool analysis as being about a *switch* has been carved into our readings, sedimented into our interpretations. Verbeek, as we've seen, already nuances the view of the presence-at-hand/readiness-to-hand dichotomy being purely about alternation, pointing out how some technologies can also be, or even *must* be, in a present-at-hand state or at least have present-at-hand aspects when in use (one of his examples is playing the piano). But this still leaves open the question of how we should understand the interaction between the two modes, and thus, the two sides of the central dichotomy. A completely different contemporary take on the tool analysis comes from Graham Harman, who with his object-oriented ontology (OOO) delivers an interesting counterpoint in this investigation.[74]

Harman's OOO as such is a grand critique of relationalism: the view, dominant in philosophy since the twentieth century, that the world is made up of relations and of nothing else—no substance or essence, as older worldviews such as Aristotelianism would have it. Harman reintroduces substance into ontology, but in an unexpected way: namely, building on the tool analysis. According to many, the tool analysis represents the hallmark of relationalism: before we conceptualize things as objects, placed before us in their abstractness, we are always already engaged in a relational totality in which things acquire meaning in practices, such as the use of tools. Or at least so the *pragmatic* interpretation would have it, Harman argues—and he goes on to turn the scheme upside down. We have been looking at the tool analysis the wrong way, Harman claims: "The hammer itself might easily be taken for something relational. But this is the central falsehood of mainstream Heidegger studies."[75] Readiness-to-hand—"tool-being," he calls it[76]—is not relation; it is substance. And tool-being constitutes an autonomous, "subterranean realm," enigmatic and unattainable. In order for it to be visible, "unconcealed," it would have to be present-at-hand, thus, relational. Only relation is perceivable. But the "real object" stays forever concealed. We can never grasp or perceive it directly. "Phenomena are only rare cases of visible things emerging from a dominant silent background of equipment."[77] This, importantly, counts for every object individually: tool-being is not one gigantic force or domain, in which all things in some or other way partake. *Every* object harbors its own substance.

Reintroducing substance for Harman does not lead to blunt materialism. On the contrary, he finds that most relationalist approaches today are avowedly materialist. His book *Immaterialism* clarifies this.[78] In it, Harman elaborates a critique of the school of "new materialism" and of Latour's ANT.[79] These approaches foreground "context, continuity, relation, materiality, and practice" as building blocks of reality.[80] Harman's central point is that things cannot be reduced to their effects, their relations, their "agency" alone. There must be a surplus somewhere, which is for him the "real object," the substantive core that every object harbors within itself. Once again, though, that core is not material in the classic materialist sense; reducing objects to a material core would be a case of what Harman calls "undermining" (reducing things "downward" to their constituent parts); while ANT and other relationalisms—and thus, the new materialisms—are guilty of "overmining" (reducing things "upward" to what they do or their effects). Harman with his real object finds a way in between.

But what could be the relevance of such notion? What is the point of introducing a substance, locked within every object, that we cannot perceive or grasp? In reading Harman's work, it might seem puzzling at first. Yet notwithstanding his aversion to pragmatic explanations, Harman continuously gives hints as to the "usefulness" of the substance concept. His answer to the question "What is the use of posing substances beyond relational grasp?" could be phrased as follows: because of the little misty "extra" that substances offer. They are a sort of ontological vermiform appendix, of which it was once thought that it serves no particular function in the human body—yet it could still burst. It could still mess up the ordinary course of events. Substances are the little "something more" that possibly disturbs every present situation. They are the "nonrelational actuality" that offers the "material" for so many significant or nonsignificant shifts of direction: "Unless the thing holds something in reserve behind its current relations, nothing would ever change."[81] Substance lurks beneath the surface, it stays needed as a subliminal "engine of change,"[82] feeding and fueling relational constellations with a sort of actualizing energy.

But taking substance on board again is just one innovation of Harman's framework. There is a second big maneuver that he makes, shaping his object-oriented ontology. This relation-substance structure is not just something that typifies humans' interaction with objects. By logical implication, it counts for the interaction between *all* objects: material, immaterial, organic, inorganic. All things in the world in Harman's view are on equal ontological footing in this regard. All objects interact with each other across this chasm of ingraspability: "Not only *human* relations with a thing reduce it to presence-at-hand, but *any relations at all*."[83] Harman often gives the following example, derived from Islamic occasionalism: when fire burns cotton, the fire meets the cotton "as" cotton, but part of the latter remains hidden to the fire, as it remains hidden to all other things—its tool-being. The fire can only grasp the cotton as present-at-hand.

And then, finally, there is a third move: Harman adds another dichotomy to this model. That makes his objects *fourfold*, namely, a constellation of two dichotomies—two "twos" crossed with each other. From Heidegger's 1919 lecture course Harman takes the ontological distinction between "being something at all" and "being something particular."[84] In another instance, he derives this second dichotomy from the work of Edmund Husserl, who distinguishes between a unified (intentional) sensual object and its plurality of traits (or qualities).[85] The sensual qualities of an object are given in experience. The unified sensual object is always encountered "through" them. This other pair is fundamental to Harman's outline of his "quadruple object." For combined, the two dichotomies—concealment versus revealment; object versus qualities—make up a fourfold representing the ontological structure of each and every thing. The second pair (object versus qualities), nevertheless, will concern us less directly here.[86] For the purposes of my investigation, what is most pivotal are the relation-substance dichotomy and the idea of this structure characterizing all things in the world and their mutual interactions. It is mainly here that we can find congruences with Bateson's perspective, which will be of further use to us.

As with Bateson, it takes some practice, some getting used to, to really *feel* Harman's point. This might be a key for unlocking Harman's OOO: like Bateson's perspective, it needs to be regarded as a prism through which one looks at the world, not simply a "theory." One must immerse oneself in it first before reaping the rewards, rather than look at it merely as a bag of ideas. Eventually, one may start to realize that Harman is really trying to convey a very simple, commonsensical insight: experience—as the generic interaction of all things with each other—can only be *selective*. A longer quote from Harman's essay collection *Bells and Whistles* helps to grasp this. In it, Harman refers again to McLuhan's "figure/ground interplay," which we have already seen (in chapter 2) to correspond with the presence-at-hand/readiness-to-hand distinction:

> How could flames and ocean waves exhaust the reality of what they strike any more than we humans can? The fire interacts with the flammability of the cotton, not its whiteness and softness; the waves strike the sand as a feeble barrier, unaware of the odors that dogs have found there. The figure/ground interplay has nothing to do with the difference between conscious and unconscious and everything to do with how things are distorted and simplified by any relation whatsoever, whether with human or nonhuman things.[87]

Now this image of things coming into contact with other things, but from the perspective of their very own "experience," resonates—quite unexpectedly, it might appear at first—with Bateson's outlook. Admittedly, Bateson's ontological "widening" isn't done as radically as Harman's.[88] Bateson's big distinction is between living things (Creatura) and nonliving things (Pleroma). But look at a statement such as this: "In the end, at the last analysis, everything we say about Pleroma is a matter of speculation."[89] Such a line could come straight out of a handbook on *speculative realism*—the movement that Harman helped to found and that wants to "speculate" on realities beyond our epistemological grasp. Indeed, Bateson's pronouncements on Pleroma at times sound a lot like Harman's declarations on substance. By implication, Creatura would then correspond with relation. Still, Bateson obviously emphasizes the relationality of the world: there is *first* relation; only in a second instance can we speak of things. But what to make of ponderings such as these, that zoom in on living versus nonliving beings but that map remarkably well onto Harman's scheme of object-object interactions?

> The things-in-themselves (the *Ding an sich*) [i.e., Pleroma], which are inaccessible to direct inquiry, have relationships among themselves comparable to those relations that obtain between them and us. They, too (even those that are alive), can have no direct experience of each other—a matter of very great significance and a necessary first postulate for any understanding of the living world.[90]

Harman also retains the thing-in-itself (*Ding an sich*) from Kant; it's his real object, though of course he will argue that the *noumenon* does not just stay inaccessible to humans, "every inanimate object is a thing-in-itself for every other as well."[91] All the more striking is the similarity at this point with Bateson. True enough, we should be mindful of not reading too much of the Harmanist scheme into Bateson's framework. The two authors are essentially working in different contexts, talking about different things. But the purpose of putting the two approaches right next to each other here is to bring out lines of force that are rather subdued in Bateson's work, and that can subsequently aid us in better modeling the central dichotomy. Bateson is not conceptualizing a "substance" completely akin to Harman's real object, evidently. Yet something in his "double description" model dovetails well with the way in which in Harman's framework substance and relation are intertwined. Remember, structure and flux are interwoven; in Mary Catherine Bateson's words: "Creatura and Pleroma . . . are not in any way separate or separable, except as levels of description."[92] What we perceive, *the only thing that we can perceive*, are *differences*—and differences are *relations*.[93] As we already saw in chapter 3, in the Batesonian framework, *not all* is relation. Let's review one more time Bateson's definition of "Epistemology" (with capital "E"): Epistemology is "the science that studies the process of knowing—the interaction of the capacity to respond to differences, on the one hand, with the material world in which those differences somehow originate, on the other."[94] Likewise, in OOO, from "out of" substance—a hidden, submerged realm—relation emerges: things that we can notice and perceive. Relation and differences "somehow originate" from a deeper realm.

Now what is interesting is that OOO provides us with a way of grasping the Batesonian perspective to the extent to which it, OOO, emphasizes that the substance-relation interplay is a matter of *each object individually*. But objects in the sense of OOO are no atomistic business. They form new objects in coming together, temporarily or permanently, with other objects. Harman in *Immaterialism* works out the case of the Dutch East India Company: a gigantic object, composed of countless other objects, but nevertheless having its own unity and autonomy, and its particular "life" trajectory. And that is just one case. We have to imagine that we—as humans, but it counts for every object—are constantly enveloped in innumerable objects this way. In assessing OOO, we might be too quickly seduced into thinking too literally of objects: clearly definable objects; something "outside"—probably once more a remnant of our Cartesian, dualistic legacy. As mentioned, to truly harvest the fruit of OOO, we need to exercise ourselves in *feeling* the perspective. "Object-oriented" may in this regard be a misleading term; it suggests orientation *toward* objects, while actually the correct topological imagery concerns orientation *from out of* objects. We have to take the inside view; however, not just merely inside, but looking out *from the inside to the outside*. And just like OOO, Bateson's framework is not a matter of an overarching, godlike perspective (mind-Creatura on this side, matter-Pleroma on that side). It is a situated stance, in which each situation, each standpoint, exhibits the characteristics that we outlined: a composition of structure and flux; taking form across different, interlacing levels of abstraction; and in which we can strike a balance between conscious-purposive striving and systemic, wise awareness of those interactions.

The Ubiquity of Breakdown

So we have to learn to take the *inside-to-outside* standpoint, to *feel like* an object in the OOO sense. This is obviously not to say we have to "objectify," in the common-sense use of the word, ourselves or others. It is about the realization that all things share a common "perspective," looked at from an OOO angle, and in a slightly different version, from a Batesonian viewpoint. The convergence even goes so far that Harman is not too intimidated to ascribe "psyche" to (real) objects, suggesting a concept of cognition that comes close to Bateson's: "All real objects are *capable* of psyche, insofar as all are capable of relation; for real objects have psyche not insofar as they *exist*, but only insofar as they relate."[95] Compare Bateson: "Mind is an organizational characteristic, not a separate 'substance.' "[96]

Of course, once again, Bateson and Harman have different priorities, and we shouldn't jumble them too much: Bateson would not assign mind, even in his broad sense, to objects. In fact, mind "needs" matter, to make distinctions in. Yet we've also seen how Harman discusses the debate on materialism within the humanities. And there is an elementary connection here. Philosophy of technology, as we saw in our discussion of the "material turn," can be said to be just as focused, especially after the empirical turn, on material practices as the perspectives that Harman discusses. Indeed, this is, in a paradoxical way, part of its proposed anti-definition of technology: technology "is not. . . ." Notwithstanding philosophy of technology's subversion of reified definitions of technology, its underlying perspective is at the same time still largely of a materialist bend. It always at least risks reverting to blunt materialism—if only because we name it philosophy of *technology*. The sheer hint of that word, no matter how much negation—*is not*—we try to import, evokes the notion of materiality, or certainly of some well-circumscribed "something." OOO, then, enables a way out of the conundrum—by offering us a pair of glasses through which to look at the world *past* the logical tension between relation and substance.

Philosophy of technology has shown how technologies beyond their thinglike appearance are actually relational constellations. But what are these relations? *More* things: actors in networks. Therefore, there remains in the end no way to delineate technology; we've lost our object of study. Indeed, this is the sort of relativism that ANT has been criticized for: ultimately in its view "all things are equal." With Harman is introduced a *difference*, a *lack*: something that always escapes—wherever we stand.

Matt Hayler has done much to put Harman's outlook to work for the study of technology, combining OOO with among others postphenomenology and embodied cognition theory.[97] Hayler defines technology, on the basis of Heidegger's tool analysis, as an "encounter"—more precisely, with an object in use. But a technology only becomes a technology when it is used with a certain level of skill: expertly. And the process of becoming an expert at using a technology entails a continuous meeting and anticipating of the unexpected aspects or events that every technology has in store, a process that Hayler conceptualizes via Harman: learning to use a technology well means gradually getting acquainted with its hidden substance.

Hayler sets us on the way to a full deployment of OOO in the service of philosophy of technology, but I believe we need to go further: his account does not fully grasp the radical consequences of the Harmanist perspective. It is not only about expert use of a "technology." Applying OOO to the study of technology is one thing, and that's fine. But we remain stuck within the definitional conundrum described above. Vice versa, we must also try to determine what the implications are of OOO *for* our understanding *of* technology. Along these lines, again, I contend that Harman's OOO stays largely unintelligible as long as one keeps imagining literal objects.[98]

A key to this is remembering once more that Harman builds his theory on a new interpretation of the *tool analysis*, and thus, of *breakdown*. Heidegger observes how we are catapulted out of readiness-to-hand when breakdown occurs: the tool we are using gets broken or lost. At that moment we abruptly need to pay attention to the tool, or the idea of the tool, in a conscious way. We start entering, or at least go into the direction of, the mode of presence-at-hand. This is Heidegger's critique of Western metaphysics in a nutshell: the objectifying attitude of metaphysics regards the tool "concept" or "substance" as the true reality, whereas in fact what comes first is the relational network in which the tool use is enfolded. Harman, we saw, turns this frame upside down. For him readiness-to-hand, "tool-being," is substance; while presence-at-hand, breakdown, is relational. "Individual objects are smothered and enslaved [i.e., by tool-being, into which they withdraw], emerging into the sun only in the moment of their breakdown."[99] Against the readings of the tool analysis in terms of merely practical tool use, Harman states unreservedly: "Heidegger tells us no more and no less than this: 'all reality has the structure of the tool and its breakdown.'"[100] So, this dynamic of something-staying-hidden is a characteristic of all the objects we are "in": *ubiquitous breakdown*. And once again, this counts for all object-object interactions. It is not exclusively us humans who have "access" via breakdown (as in what Harman calls the "philosophies of access," that is, human access, that arose in Kant's slipstream). All things partake in this dynamic.

The consequences for philosophy of technology are elemental, and twofold. OOO invites us to take the inside-to-outside view, but (1) not *only* from within "technology" (that is, what we commonly know as technology), given that all objects have equal ontological status. Everything is a technological interaction, or better, technology takes part in an interaction that typifies all objects. Yet (2) this is the case for objects *among each other* just as much; an insight that is really starting to become tangible nowadays in times of emergent globe-spanning algorithmic constellations (Wiltse's "mediating (infra)structures") that more and more escape our conscious human intervention. Most crucially, notice the analogy between Harman's relation-substance distinction and the central dichotomy. The key insight that Harman delivers here is that these modes are two sides of the same coin. Philosophy of technology, by contrast, tends to choose sides: technology is only seemingly about the narrow mode and *really* about the wider mode. Yet why not consider this: *at all times both are at work*. The way we meet things, at first in the narrow mode, is through relation/breakdown. This is where we cast things as just efficient elements in our doings and activities, our "in-order-to's"—often without much conscious deliberation. By definition, something always stays concealed. "Our use of the floor as 'equipment for standing' makes no contact with the abundance of extra qualities that dogs or mosquitoes might be able to detect."[101]

I can illustrate further with a metaphor. Imagine the dynamic I describe as moving through the world in *circles*.[102] These are environments: fluently we maneuver within them and through them; our acquaintance with them needs no deliberation. The carpenter, for instance, knows the ins and outs of his trade. The academic scholar swiftly navigates her domain. These realms have their very own purposive structures, undoubtedly and inevitably in the sense of instrumental purposiveness (narrow side)—the scholar, for example, needs to publish in order to make her ideas and work known; and as instrumental purposiveness typically tends to hypertrophy into perversion, in the academic climate of today she is pushed to publish as much as possible without much regard for quality and content—but certainly also in a wider-purposive sense as well: both carpenter and scholar "know how." They do "their thing," meaningfully relating to a world that they partly co-create in doing what they do. Ideally, they have Stiegler's *savoir-faire* and *savoir-vivre*; and they take part in Borgmann's focality.

But still the two, carpenter and scholar, are to an extent worlds apart. Come into contact with each other, they'll quizzically look out from their own circle, faintly informed about the other's occupation. The general terms in which they refer to the other's business will strike the latter as crude. Inside the circle, outside the circle: those are incommensurable realities. Carpenter and scholar may both be avid squash players. Another circle. Circles can overlap, fold into each other. And we now see the two embarking on a fiery conversation about their sports, needing only cues and hints to understand each other. Indeed, the circles' membranes are not completely impermeable. Ask the carpenter and scholar to switch professions: both would have to painstakingly familiarize themselves with the new world—like learning to use a strange tool. But they may succeed. They might start to grasp each other's *purposes*. Outside may become inside. The unfamiliar familiar. The foreign, home. But still then, so many other realms will remain out there, unexplored.

This way, we look out, from inside the "object"—the environment in which we fluently move (my trade as a carpenter)—and something always escapes. Our grasp of other objects, other circles (the world of the academic scholar), remains incomplete, even though we may explore and discover other objects gradually; learn a new trade for example. The idea of a *gap* is central here. Both Bateson and Harman stress the necessity of some gap between the two big realms that they see interacting (Creatura/relation, Pleroma/substance). Harman argues that the main gap in ontology does not lie between the human being and the world, as philosophy since Kant would have it (with the former gaining some access, however failing to gain full access, to the latter). No, the gap is situated, on the level of *all* objects, between relation and substance: "the basic rift in the cosmos lies between objects and relations in general. . . . Whatever the special features of plants, fungi, animals, and humans may be, they are simply complex forms of the gap between objects and relations."[103] Bateson in turn, as we saw in chapter 3, speaks of gaps in the structure-flux scheme, mentioning the "necessary incompleteness of all description, all injunction, and all structure."[104] Gaps understood in this way are a characteristic of Creatura; they do not exist as such in Pleroma. And "however much structure is added, however minutely detailed our specifications, there are always gaps."[105]

Now the funny thing is that, exactly because of the broadening of ontological focus that OOO offers—*all* things are at stake—we retrieve, quite inadvertently, philosophy of technology's object of study. On one hand, logically thinking through the analogy, the efficiency-oriented versus impact-aware dynamic sticks to *all* things. On the other hand, technology can be said to be exactly this "circles" structure of knowing-well-on-the-inside versus staying oblivious of an external, but possibly shifting and morphing, outside. This also means that philosophy of technology is not about technology in the strict reified sense—it is *philosophy*.[106] By losing our central object of study, we find it back. We are at every time in a technology. And extending Hayler, we could say that an "expertise" structure of sorts characterizes all spheres of life.[107] The world is like technology well used. It *is* technology well used (and obviously sometimes *not* well used). Vice versa, technology takes part in this all-encompassing dynamic of inside and outside—this means, of narrow and wider purposes. We are at any time on the inside looking out. Out may become in. But then *another* "out" pops up somewhere—always. We are moving through life, navigating circles that we master, while looking outside to the lesser known. We may get better, but things never become all there. Breakdown is ubiquitous. And it is ubiquitous in the way that it isn't really a switch—or at least not as we commonly imagine a switch—between the two modes of the central dichotomy; the two modes are always there at the same time.

In this view, of course, *context* is central. Mary Catherine Bateson in *Angels Fear* advises: "Within a given sphere of discourse, strive for the consistency that fits the logic of that sphere."[108] This, we may add, does not only count for discourse. It counts for "material" constellations too; I will return to this issue right away in the following and final chapter. Technology is both about using and talking "correctly." But in fact, in this sense, the reverse is the case just as much: using and talking is, philosophically speaking, a matter of technology, that is, of purpose. We move toward a perspective in which we regard technology not so much as *this* or *that*, but in terms of a *here* and *there*. Phenomenologically, we are in the experiential standpoint first of all. But this experience also involves, if only indirectly, the nonexperienced: what is not experienced, what goes on "behind our backs." Only if we get rid of our unstoppable inclination to make things appear—in a conceptual as well as practical-technical sense: chase after our conscious purposes—will we be able to create some distance between us and the so-called thing that technology is, or actually *is not*, and acquire some ecological sensitivity to the network-like character of our technologically mediated world. This should be the cornerstone of a program for an *art of living with, in, and through technology*.

CHAPTER 7

The Art of Living with Technology

There is only one way out if we want to think *disappearing technology*. The organic conclusion of all the foregoing would be that we have to think *ourselves* technology, but not in the commonly accepted sense of "we humans are technological beings." We are technology to the extent that we purposively act, think, and cope in a world from out of the object—in the Harmanist sense—that we are. Looking out from *here* to *there*, we always partake in the two modes. Over there are our narrow purposes, standing before us reified, objectified: aims we strive toward. Over here is, generally less conspicuous, the pool we swim in, of relations, network structures, that have a purposive structure as well, but of a much wider, noninstrumental, more elusive variety.

This topography still presumes that we humans are technological, ontologically. But it also hands us a tool to reflect on ontic fusions with technology. Those are not the fusions as we have long tended to understand them: additions of one component or entity (technology) to another one (human being). We are already fused to that extent. We are already *circles* interacting, intertwined with countless other circles. *That* we intermingle, as object with other objects, is a given. But at least we can try to get some grasp on that intermingling. Indeed, "we can *try*" and "*some* grasp": these attempts must always be incomplete and temporary. Breakdown is ubiquitous: something always escapes. Yet we can learn to *see double*, to see ourselves as wrapped up in this "two" structure—of the two modes enveloped into "one"; looking out from *Zuhandenheit* to *Vorhandenheit*. As Bateson puts it: "Two descriptions are better than one."[1] Hence we become acquainted with a portrait of ourselves as a being that's aiming-toward-something, but within and from within a context. This bridge between two spheres is what "technology" is about. As far as it is what technology is about, it is what we are about—and what many other things are about, for that matter.

But then the question pops up: how to *live*? How to live with, in, through technology like this, seen through this lens? In this final chapter, I go into this question, tying up the loose ends of my investigation. This will imply revisiting one more time a few of the central ideas in the Batesonian conceptual catalogue, expanding upon them and resituating them in the philosophy of technology framework. The final goal of this endeavor will be to leave the reader with a way of recontextualizing and perhaps offsetting the schizophrenic view that typifies our contemporary cultural evaluation of technology. Once again: we tend to be instrumentalists (utopians) in our everyday practice and routines, and determinists (dystopians) in our cultural imagination. Strangely, this appears like a sort of reversal—dreamlike and nightmarish at the same time—of our two-mode scheme. We are looking out at images of dystopian gloom and widespread disaster ("everything will go wrong") from within the safe haven of our instrumentalist, narrow-purposive expectation ("all will be as planned"). The actual situation is that we can only do our best to act, as conscientiously as possible, upon our conscious purposes from within, from out of an unoverseeable network-of-impacts. Some things will go as planned, but definitely a lot won't—better start accounting for that, learn to live with that *now*.

Yet perhaps we do not need to *cast out* the schizophrenic attitude. On the contrary, we might have to embrace it. For Bateson, we saw, schizophrenics always perceive "more," engaging as they do with multiple levels of abstraction (be it, seen from his framework, mostly in a way that creates pathology). We can learn something from them, learn to see "more" as well. The art of living with technology seeks to practice aspects of this purported madness. Seeing double is seeing more—that is, even more than the "more" already mapped by the wider side of the central dichotomy, because we acquire an active epistemological engagement with the *interrelation* between the narrow and wider side. And luckily, there are some practical cues available to set us on our way—albeit that their practicality might not be of the sort that we've come to expect, craving for "concrete" solutions, measures, or recipes. My cues can only frustrate those cravings. And it is only to the degree that they frustrate, that they can be in any way—if that term applies—successful. I won't deliver a clear-cut recipe for the "art." The art needs to be practiced, in every new context. But at least I can, with the components that we've gathered so far, draw the outlines of a program that leaves enough leeway for variation but still puts us in a distinct direction, if only by changing the terms of debate (to begin with in the philosophy of technology).

In a first step, I will revisit the topic of disappearing technology, which will serve as the springboard for a further reflection on invisibility, coupled to the phenomenon and technological development of algorithmization. The Batesonian notion of levels of abstraction can help us to make sense of these issues; and to elaborate this fully, we must look again, yet more closely, at the conceptual pair of the abstract and the concrete, and how these two interrelate. This will bring me in a second step to the related pair of discourse and matter, a well-known conceptual couple in the philosophy of technology, that I already discussed a few times, but that now needs reevaluation in the light of the framework I've been building up. Spontaneously, then, and at last, this new evaluation will make us arrive in the midst of crucial questions about efficiency. "Efficiency" as we commonly know it *and* as it is used and conceptualized in the philosophy of technology needs reevaluation as well, seen from my framework. Thus the third section seeks to break open the notion of efficiency, from, among others, historical and sociological points of view, in order to lay the groundwork for the exercise of the fourth section. That exercise involves resituating the remodeled notion of efficiency in the philosophy of technology—more precisely, resituating it within the central dichotomy. All these elements, which are of course essentially connected to *purpose*, form building blocks for the program of an art of living with technology as proposed here. For further guidelines on how to "apply" these—here too, as far as this perhaps too-strategically oriented terminology is warranted—I turn in the last part of the chapter one more time to Bateson's cultural critique, highlighting especially his reflections on quantity and quality, the analog and the digital, art, and coping through contexts. Capping off the edifice, these elaborations make it clear that the key to the art is one of the hardest things to do, namely: reexamining our epistemological presuppositions.

Technologies Disappear: A Matter of Levels?

So what do we *do* with disappearing technologies? While more arcane "invisible" technologies such as nanotech are still relatively removed from our daily doings (for the time being), things like algorithms, artificial intelligence, and machine learning are smoothly oozing into everyday life. A world emerges in which formerly "dumb" objects become "smart" and full-blown collaborators in human schemes—think of phenomena such as Internet of Things, informatization, and digitalization of more and more domains of life, widespread implementation of sensor technologies, ever-growing adoption of mobile devices and social media, and so on. *Algorithmization* is affecting almost every realm of human practice.

At times, algorithmic (infra)structures even appear to be taking on a life of their own, beyond conscious human intervention. Examples range from the spectacular and enigmatic to the everyday and (almost) mundane. During the "flash crash" of May 6, 2010, for about half an hour the American stock market, grounded heavily in algorithm-driven high-frequency trading,[2] crashed for mysterious reasons—signaling, as would become clear, the degree to which algorithms and no longer so much human intention steer the stock exchange. More than simply an economic problem, it indicated that algorithms' rapid-fire interaction has outgrown human control.[3] Closer to home, Google's Photos app automatically generates a photo album, for instance of one's summer holiday, on the basis of image processing and data from one's smartphone location. For most people it probably still comes as a surprise to suddenly find a computer-compiled photo collection of "Your trip to Manchester" on one's smartphone; but how long will it take before we accept this as the new "normal"?

To an extent, what we go through is really nothing new. The invisible has always intrigued, puzzled, and haunted humankind. From the imperceptible come threats and danger, but also the surprising and the unexpected, in more positive senses. Things we don't understand must have some explanation in a realm that we cannot pierce through to. Gods, magical forces, physical laws: invariably they do their work from out of a "place" we cannot perceive, or perceive yet. In a way the history of Western philosophy can be understood as a struggle with the elementary tension between what we see and don't see. Admittedly, in recent centuries, due to the world getting more and more complex, the invisible has grown in our lives. It has been "everydaylized": life is in so many aspects tinged with traces of the unperceivable. Our food arrives from invisible places—far-removed fields and factories, distribution networks, packaging and marketing industries. We only perceive the neatly wrapped, tightly priced piece of meat in the grocery store rack. The penicillin that magically cures our infection is produced somewhere in a lab by way of procedures we don't fathom; we only see the pill. One could ask: as far as invisibility is concerned, to which extent is the magical antibiotics pill—coming to us from mysterious medicinal realms—any different from the shaman conjuring up good spirits in order to heal the sick? Or how far removed from the abstract conditions of industrial food production was the uncontrollability in hunter-gatherer times of weather and ecological factors? One could surmise there exists something like a baseline of invisibility, of ungraspability.

Yet today, we might say, something more is going on with invisibility—to the extent that never before has invisibility been so *designed* into the fabric of human existence and sociality. To an everyday user of ICTs (information and communication technologies), much of the networks of data, artifacts, and other components that constitute them remains unnoticed. That would not be very pertinent, were it not for the fact that we can presuppose these behind-the-scenes structures and networks to have all sorts of *effects* on us users. These effects are sometimes intended by the algorithms' makers—as for instance in the case of persuasive technologies such as smart toothbrushes or waste bins, that push us gently toward the "right" behavior. But even unintended consequences might be "harvested" in a research-and-development way for future exploitation: algorithm-steered media services such as Netflix and most social media try to construct profiles of their users on the basis of what the latter click on, like, post, et cetera, to be used later on for purposes of targeted advertising. For the mechanisms (code) that have this steering or controlling impact to stay hidden is obviously in the interest of the makers.

This second, intentional invisibility makes our predicament even more acute. We have to deal with technologies disappearing; that is one thing. But some people, companies, governments, organizations, and so on *have stakes* in the disappearance; that is another issue. To us in everyday practice, however, the result is largely the same: disappearance "appears" as disappearance. That's that. Gradually we might get accustomed to this elusive given, always in the background, that so much is going on behind things' façade that we don't see or cannot see, and perhaps eventually we will not even want to ponder it anymore. As research in media studies tends to demonstrate, this is already the reality for many people: they just don't seem to care that much about privacy, for instance, unless they are directly prodded and questioned about it. But daily practice quickly forgets conscious concerns: terms of service are accepted unthinkingly with the installation of every new app.

Yet do we have to counter this indifference by the reverse attitude: paranoia? The systemic wisdom that Bateson advocates might in its most extreme form indeed come close to something like paranoid suspicion: there is always something "more." There is always something hidden behind what we see.

This seems like a deadlock, but the way out, as usual, is to truly attempt putting on the Batesonian glasses and looking at the issues through them. We have to teach ourselves to see ourselves differently. This is very hard; as we've seen, it takes distancing from our whole intellectual-philosophical legacy. And it must turn out that, at the end of our conceptual travels here, even the imagery of *disappearance* is partly misguided—steeped as it still is in contrapositioning an overarching "us" and an overarching "them." Acquiring a new vocabulary of course doesn't happen overnight; we still have to make do with the old words. Even dyed-in-the-wool contemporary philosophers of technology such as Mark Coeckelbergh, who speaks of "the end of the machine," referring to the given that ICTs seamlessly fuse with our everyday being, still relapses now and then into all-too-dualistic imagery (but my text here is just as guilty). "We no longer see computers and related technological realities as machines," Coeckelbergh observes.[4] The point, however, is not only that technologies are disappearing from view. They *are* of course, seen from an everyday perspective. But they are as a matter of fact also *disappearing—as we know them*. "It's the end of technology as we know it," with a variation upon the famous R.E.M. song. This is the logical conclusion of the (ontological) human-technology continuity. Technology simply does not exist; it cannot exist. But still we know somehow, feel somehow that we deal with technology, or what we formerly knew as technology, and that we are surrounded by it—indeed across a spectrum from visibility to invisibility. So what are we talking about then? We are talking about purposive structures, in which some things appear and others disappear. And the crux of the matter might be that it wholly *depends on the standpoint from which one looks* whether things appear or disappear. This seems like a trivial observation, but it is crucial.

Depending on which standpoint one takes, from out of which circle one looks—whether real or imaginary—a certain ratio between visible and invisible comes to the fore. At this point Bateson's levels of abstraction approach can be seen to dovetail well with aspects of Luciano Floridi's philosophy of information. Floridi also adopts the notion of levels of abstraction, albeit that he doesn't derive it—via logical types—from Russell, but from computer science.[5] He defines a level of abstraction (LoA), in a paper written together with J. W. Sanders, as "a collection of observables, each with a well-defined possible set of values or outcomes."[6] An "interface," then, "consists of a collection of LoAs and is used in analysing a system from varying points of view or at varying LoAs."[7] LoAs are not necessarily ordered in a sequence; they might be "nested, disjoint, or overlapping and may be, but do not have to be, hierarchically related or ordered in some scale of priority."[8] But what does a LoA do? It "qualifies the level at which a system is considered."[9] Floridi illustrates by way of the example of a group of friends at a party talking about a motorized vehicle. One is a collector, another is an economist, a third person tinkers as a hobby. Each talks about "the system under discussion" from a different perspective, emphasizing for instance the general state of the vehicle, its market value or the type of engine, respectively. "Each LoA . . . makes possible a determinate analysis of the system."[10] The outcome of that analysis Floridi calls a "model" of the system.

Now one might ask what is so special about these observations; people always view things from (out of) their perspective. Yet Floridi introduces the notion of LoAs in order to clean up confused arguments and discussions. LoAs are a method of straightening out the different strands in a debate or the different lines of force in a problem. "In general, it seems that many uninteresting disagreements might be clarified, if the various interlocutors could make their LoAs explicit and precise."[11] It's not just about trivial matters, though, if one would be inclined to think that. "Too often, philosophical debates seem to be caused by a misconception of the LoA at which the questions should be addressed."[12] Floridi adds that "this is not to say that the method represents a panacea," but surely it is a powerful instrument. Floridi himself applies it to, among others, questions of moral agency of artificial agents and of ethics.[13] In each problem, one has to try to identify the levels of abstraction at stake and see which "observables" are situated within which LoA. Importantly, the choice for a LoA, Floridi points out, is driven by the *purpose* of the "original request for information."[14] "LoAs are teleologically oriented."[15]

Floridi speaks a different language than Bateson—if only with regard to the bigger theory in which they are located. Floridi's definition differs from Bateson's perspective. But still the former can help to illuminate the usefulness of the latter. Like Bateson, Floridi seeks to crystalize a from-the-inside view. In treating of certain problems, we might be looking at them or aspects of them at the wrong LoA. With Bateson, we might misunderstand phenomena if we don't account sufficiently for *context*—thus, also, for the level of abstraction at which certain relations take shape (notwithstanding, again, Bateson's and Floridi's own different contexts). This sheds another light on problems with regard to invisible, disappearing digital (infra)structures. We just might have been looking at them at the wrong level of abstraction, namely, from out of our construed standpoint as observing "subject." Indeed, from that point of view, these structures or things *can only appear as disappeared*, as invisible. At the same time, so many actors behind the scenes are able to put this to good commercial or political use. The result that we get, then, within this perspective, is indeed a form of paranoia: "I cannot see things, but I know/feel they're there." With regard to ICT, this is nowadays a common cultural reaction. "Who knows what kind of powers lurk beneath the surface, behind the curtains. Who knows who knows what or sees what." But paranoia, epistemologically speaking, is only possible when one's view remains restricted to just *one angle*. The paranoid attitude is constructed around a duality: *they* are after *me*. The fact of "them being after me" might be true or not—this is what the quip "Just because you're paranoid doesn't mean they aren't after you" conveys.[16] Paranoia may just be warranted, may be a legitimate attitude. The thing is: one doesn't know—there's only *one* perspective to judge the situation from, namely, *mine*. With the exercise of conceptualizing, taking on board other levels of abstraction—if only as a thinking experiment or as an exercise in imagination—we can practice ourselves in taking other standpoints. Instead of only expecting, guessing at, or fearing "the other things," we can attempt to visualize them, map them. This should not mean that we need to cast aside—or worse, repress—our paranoia. Perhaps we should embrace it—just like there is, in Bateson's view, something about schizophrenia that we should embrace. We should look for a beyond-paranoia. Paranoia might as well be the starting point for exploring other perspectives, other LoAs. It might be an entryway into the exploration of—putting it in a Marcusian way—the abstract conditions of the immediately given-to-us concrete.

Nevertheless, there is some ambiguity here. The abstract and the concrete are nowadays intertwined in even more intricate and confusing ways than Marcuse could analyze in his day. What do *ICTs do* with the concrete and abstract? Douglas Rushkoff frames digital technologies as "biased towards abstraction," as they are "bringing everything up and out to the same universal level."[17] In the digital realm, scaling is the norm. The Internet companies that are able to climb up levels of abstraction—becoming search engines, aggregators, or portals instead of doing business themselves, for example, providing information[18]—are the most successful, Rushkoff claims. But it is not just a matter of economy: there is something structurally characteristic about digital technology that makes it inclined toward the abstract. "In a universe of words where the laws of hypertext are truly in effect, anything can link to anything else. Or, in other words, everything is everything—the ultimate abstraction."[19] At the same time, Rushkoff suggests that young people are becoming proficient in intuitively grasping the essence of a topic by just briefly interacting with it: "For them, hearing a few lines of T. S. Eliot, seeing one geometric proof, or looking at a picture of an African mask leaves them with a real, albeit oversimplified, impression of the world from which it comes."[20]

This is reminiscent of some of Michel Serres's observations in his much-discussed *Thumbelina*.[21] But Serres evaluates the concrete-abstract ratio differently. His essay sings the praise of the digital native, whom he dubs "Thumbelina" for reasons of her impressive skill in thumb-typing on her mobile phone. Thumbelina stands for a radical breach with the previous generation. She finds herself completely estranged from traditional-modern worldviews, educational models, and political-societal ideals. In that capacity she is also nothing less than the harbinger of a possible new era of human reason, Serres claims. And this is all thanks to digital media, the use of which Thumbelina is so intensively wrapped up in. Serres expects the end of "the era of knowledge," to be replaced by something else: "the new genius, the inventive intelligence, an authentic cognitive subjectivity."[22] And he goes to sufficient lengths to entrap that new kind of human reason, which ICTs are helping to make possible. Unlike traditional knowledge, the new intelligence is not divided into specialisms. It is based on creativity, not on referral to conceptual categories. In essence: it prefers concreteness over abstraction. If we ask Thumbelina about "Beauty," Serres suggests, she will start to enumerate: a beautiful woman, a beautiful horse, a beautiful sunrise. Instead of what we would expect, namely, an abstract conceptual definition, she gives us, like Plato's Meno, a list of examples. But why not, Serres asks; do we even still need abstraction? "Our machines now scroll so quickly that they are able to count particularities indefinitely—they know how to stop at originality. . . . Abstraction can be replaced, at least sometimes, by a search engine."[23]

This inclination toward the concrete, the specific, the singular, and the contingent, toward examples and stories, is what marks Thumbelina's new cognitive modality. And Serres welcomes it, as the urgently needed corrective to the older generation's tyrannical predilection for abstractions and concepts, for declarations and analyses. We may also be reminded here of Michel Puech's critique, albeit on a more general level, of discourse in favor of action (see chapter 2). One needs to undertake concrete deeds, Puech insists, instead of merely protesting in abstraction. This is convergent with (the many) critiques of the rational character of modern thinking. Morris Berman, for one, points out how with the Enlightenment, abstract, rational, conceptual forms of knowledge became dominant and highly prized, to the disregard of visceral knowledge and experience. But the latter are actually primordial, and we've forgotten that. To be sure, we *need* conceptual thinking, Berman notes; otherwise we wouldn't be able to "construct a science, nor any model of reality."[24] However, we have allocated to the abstract the status of the " 'really' real,"[25] thus confusing, in the sense of Korzybski, map and territory, forgetting that our network of concepts is merely *a model*. And, Berman adds, "this confusion of map with territory is what we have called nonparticipating consciousness."[26]

These arguments make sense, yet at the same time we should be cautious about calls to reinstate the importance of the concrete—that they not overcompensate for an emphasis on the abstract. Such pleas might sound very attractive, all too attractive, actually, in times in which the concrete, concrete results, concrete solutions, and so on are exalted. But we would run the risk of losing sight of how important the abstract in our world is. In the field of sociology and cultural studies, scholars such as Daniel Bell, Nico Stehr, and Frank Webster have proposed abstract, theoretical knowledge as nothing less than the foundation or core of modern society.[27] In previous times, innovations and technical breakthroughs were more the result of tinkering and practical experimentation, often done by people with no training in fundamental science. Webster gives the examples of Abraham Darby's blast furnace, George Stephenson's railway locomotive, and James Watt's steam engines.[28] Nowadays, by contrast, research and development, urban planning, even politics and social policies are carried out and implemented on the basis of highly abstract, systematized, and codified theoretical principles. We erect buildings, take in our food, manage our health, and much more according to sets of specialized knowledge derived from empirical research, statistics, and theory-making. Out of the abstract, so to speak, the concrete is sculpted. It would be a mistake to cast aside any engagement with the abstract as old-fashioned, politically moot, or epistemologically flawed.

Ideally, we would look at the places in which concrete and abstract are intertwined. It is actually at their intersections that essential things are happening; we need to get involved in those spots. This entails getting a feel for what both "do." As we can learn from Marcuse, any discourse glorifying the concrete to the ruin of the abstract runs the risk of, first, concealing its own abstractness and, second, more dangerously, obscuring the fact that the abstract has in any case not been done away with. The abstract, the discursive, is still there, albeit maybe in material-technological form, as attested to by Feenberg. On the other hand, overemphasizing the role of abstract knowledge and concepts may blind us to possibilities for influencing or participating in that abstract realm. Those endeavors would unavoidably have to take the shape of concrete activities in the first instance. The abstract and the concrete define each other. In that vein, the aforementioned knowledge society theorists, in demonstrating the political stature of theoretical knowledge, point to opportunities for becoming better aware of how concrete decisions and events impact upon encompassing abstract-conceptual frames and directives. This also rings through in Rushkoff's analysis, and to a certain extent in Serres's (notwithstanding the latter's apparent rhetoric *pro* the concrete). Rushkoff warns that there is still a (concrete) reality—"the here and now"[29]—and that digital technologies risk to estrange us from it by their sheer abstraction. His central counsel "program or be programmed" gets formulated in line with that: we need to learn to program or at least learn something about the workings of code—so central to the information infrastructures that determine a good part of our lives. Coding indeed makes for a very concrete undertaking. Something similar is proposed by McKenzie Wark when he avers that *hacking* starts with an "abstraction": the production of "new concepts, new perceptions, new sensations, hacked out of raw data. Whatever code we hack, be it programming language, poetic language, math or music, curves or colorings, we are the abstracters of new worlds."[30] Serres, on his part, advocates taking account of a new kind of abstraction, namely, algorithmic or "procedural"-cognitive abstractions. These stand in contrast to the "declarative" abstractions of the old epistemology; we cannot understand the new abstraction with our old conceptual tools. Remember, both Serres and Rushkoff see young people fluently taking hints from just fragments of knowledge, extrapolating from the concrete. But their shared advice—if we may synthesize it like that—is not to stop there, and learn to chart how the purported concrete is co-constituted together with a surmised abstract.

With Bateson (and to some extent Floridi) we come to realize that this relation is not a one-to-one dynamic. Rather, concrete-abstract relations take the shape of a web through which we proceed stepwise, moving through levels of abstraction—each time the view changing with the new perspective that we take, the new circle we're in. By definition, one can be in only one place at a time. This is a trivial observation—but only seemingly. We have at any time the immediately given concrete to deal with. But that concrete is always the result of abstract conditions coming to us "from afar." One could spend the day reflecting upon the abstractness that gives shape to one's concreteness. The morning radio wakes you up: a very concrete given. But its origins are abstract—at that time, for you: the radio station where the signal originates, the antennas transmitting it, the electric grid that delivers current to your clock radio. However, you would be able to go on an investigation and start mapping this network—Latourian style—and then this morning-time abstract would gradually become a new concrete, *with yet new abstract conditions shaping your new concrete.* You could do the same for the food of your breakfast. Which systems hide behind the simple-instrumental appearance of the eggs on your plate? You could go and outline the roaming area that the chickens that laid your eggs had at their disposal; whether they were administered antibiotics; which market mechanisms play out between producer, distributor, and consumer; et cetera. In this case, too, you would get a view on the abstract systems that *make* your concrete, thereby turning the former into something more concrete, but never being able to "concretize" *everything*—there is always the circle structure: one looks out from *here* to *there*.

This, now, comes to be an essential part of the art of living with "technology": acquiring an epistemological angle from which one, each time, can turn around an initial—be it spontaneous, ingrained, or learned—concern with the concrete into an involvement with the wider social-economic, political, ecological networks: the always-present, always-influential horizon against which our everyday doings play out. That is also the approach proposed—albeit not formulated in the terms I use here—by the German sociologist Harald Welzer in his *Selbst denken: Eine Anleitung zum Widerstand*, a black-humorous critique of contemporary capitalism and a rallying cry to stop the latter's endless exhaustion of natural resources.[31] Welzer argues for a form of personal responsibility, as starting point for "resistance." Obviously, there is no such thing as a socially isolated "I." The responsibility encompasses by definition other people, other species, the environment, the planet. Welzer, too, regards abstraction as characteristic of our age. Abstraction—like Marcuse argues—severs products from their functions. We buy a "pure" can of tuna in the grocery store, without wondering about how it was brought there, and with which ecological and ethical consequences. Our life has become history-less and context-less, Welzer avers.[32] We are used to outsourcing our living conditions to abstractions—in other words, the abstract networks. It takes, with a term he adopts from Günther Anders, "moral imagination" to reconnect our concrete deeds—buying tuna—with the abstract goings-on behind the scenes—such as the extinction of certain tuna species. Welzer as well makes a plea for an "art of living" in this sense, one that is always conditional, moreover: he calls it "essayistic" living—literally: probing, experimental.

With Bateson, and generally from the perspective of my framework, we can better understand the nature and dynamic of such probing. It is not a linear undertaking. With every concrete, a new abstract emerges—also in the case of invisible digital technologies and ICTs. We might be inclined to shrug our shoulders and just try to "live" with (algorithmic) invisibility—and in itself, something can be said for such a stance, as we *always* face *an* invisibility. There is no such thing as absolute transparency; think Harman. But with our approach, in the face of this invisibility, we do not need to revert to an attitude of resignation. On the contrary, the "glasses" offered by our perspective make a new appreciation of the invisible possible. Moreover, we can start to make sense of the interesting tension between on the one hand structures that are beyond conscious awareness and on the other hand *designed* invisibility, as outlined above, by way of the conscious purpose versus wider systemic awareness scheme. We need to start grasping these dynamics, namely, of what is deliberate and intentional versus what fundamentally escapes consciousness and intentionality; where do the two meet, and how? If we can find these places—these crossing points, boundary lines between contexts—and point our finger to them, we will make progress in our understanding of algorithmization. Nevertheless, this is also to make clear that a one-sided critique of algorithmization—that would only account for what it is clouding and concealing—is not what we are after. Algorithmic technologies open up and reveal much, as well: data and information that was unavailable, hidden, or in any case very hard to come by in the recent past. Think simply of the spate of information to be found on Wikipedia or Google Earth. It is not a matter of us against the Big Invisible out there. This is what our framework helps to point out: at any time, we are enveloped in a visible-invisible structure. With luck, however, we can *move*, across levels, and take another standpoint.

Discourse and Matter

But note: this across-level perspective is not symmetric. It is not that we gain an uncluttered, pure view on another level from out of "this" level. That is what my mapping of the central dichotomy onto Bateson's framework and

Harman's OOO should evince. The central dichotomy, in one common reading, clarifies how we understand technology first of all *discursively*, as instantiation of efficiency, but a closer look should reveal that *materially*, technology is much more than that: a network structure of components, factors, impacts. This reading, however, harbors an implicit idea of separation between discourse and matter (or action) as two different things, between which we alternate or (have to) choose. Obviously, they *are* different things. But they belong to each other within one and the same frame. Some approaches in contemporary philosophy of technology make this more clear than others, and it is generally something of a shrouded insight—remaining elusive largely because of the history of the field, that is marked by definitional maneuvers such as the empirical turn and the material turn.

We've seen how the material turn entailed a reaction against decades of linguistic approaches across the humanities and social sciences. Material turnists took a clear position: let's look at material practices finally, or at least *also* look at them—instead of obsessively zooming in on symbolic exchange, language, and so on. For instance, in the work of Andrew Pickering, human agency can be seen to interact with "material agency," and Pickering charts this interaction specifically in the activity of science-making, by way of his notions of the "mangle of practice" and the "dance of agency."[33] In Latour's work as well, a kindred orientation has been present pretty much since the beginning. The philosophy of technology has picked up on these cues or had parallel developments, framing technology very consciously as the "material component" missing from many discussions on human meaning-making. Ihde has long been one of the fiercest defenders of a focus on materiality, in opposition to the linguistic turn in philosophy generally. As Verbeek puts it, the material turn in philosophy of technology means to make clear that "we also need to study *things*, rather than merely focusing on *humans*."[34] In postphenomenology's terms, technology mediates human practices, and this mediation, in its material form, has often been forgotten in the investigation of those practices. As such, the material turn has offered a necessary correction to the all-too-one-sided linguistic approaches that reigned in intellectual history throughout the twentieth century.

Nowadays, however, the material turn is put more and more under scrutiny by philosophers of technology, even by some of them who initially helped to establish it, exactly because it might have overcompensated (like all turns are in risk of doing; see below) a bit too much for what it meant to correct. Verbeek, for one, makes a plea for "one more turn after the empirical turn." Postphenomenology might have understood the aforementioned mediation too much in material terms; we need to give again more attention to its meaning for humans. Material turn approaches, according to Verbeek, risk to under-investigate "the ways in which human beings *appropriate* technological mediations": "In order to understand how technologies mediate knowledge, morality and metaphysics, we should not only study technologies, but also the ways in which human beings give meaning to their mediating roles."[35] Mark Coeckelbergh has sought in recent publications to complement philosophy of technology's emphasis on materiality with a focus on discourses about technology and a study of the intersections between language and technology. He questions "the empirical turn's one-sided if not obsessive focus on the artefact, the thing, the object."[36] Philosophers of technology "have taken a material turn" under the influence of social constructivism, but they have adopted the latter "in a way that neglects language."[37] Latour himself, in fact, as Coeckelbergh points out, has given more due to language than for instance postphenomenology, influenced as he is by semiotics (indeed, he analyzes technological practices in terms of scripts, translation, et cetera). Postphenomenology, according to Coeckelbergh, has overemphasized the material; a critique in fact in line with Verbeek's own aforementioned assessment of the material turn. Coeckelbergh wants to rebalance language and materiality, stressing that both should be taken into account in the study of how we humans cope in the world. "Both words and things make us see and do differently; both language and technology mediate our relation to the environment."[38] This rings through in his wide-ranging analysis of our deeply rooted cultural ideas and discourses about technology in *New Romantic Cyborgs*. Those discourses are for a good part essentially *romantic* in spirit, Coeckelbergh argues, notwithstanding that romanticism is perhaps one of the last worldviews we would connect to technology in common-sense thinking. And philosophy of technology has also remained blind to this, because of its focus on artifacts and things, making it disregard "discourses about technology . . . and the 'symbolic' cultural significance of artifacts."[39] Coeckelbergh in response wants to bring a "cultural angle to philosophy of technology."[40]

A more "meta" approach comes from Dominic Smith, who actually asks questions about all this "turning" in the philosophy of technology, speaking of "many turns" and the "profusion of turns," and scrutinizing the underlying assumptions—conceptual as well as methodological—defining the current state of the field.[41] For him there is a structural relation to be found between the turning dynamic and the way in which the field has engaged, or neglected to engage, with the history of continental philosophy under the impulse of the empirical turn. Smith on his part, as we saw, finds a renewed engagement with the transcendental—as the conditions of technology—to be a prerequisite for an accurate assessment of technology.[42] This, too, implies a widening of philosophy of technology's scope beyond the merely material focus.

As far as these approaches constitute important reflections on what philosophy of technology should do and how it should do it, they also threaten to preserve a kind of *distinction* between discourse and matter—if some of them don't preserve it outright.[43] Other approaches, not strictly hailing from philosophy of technology, have more deliberately sought to *tie up* discourse and matter, outline them not so much as perspectives that we could or should choose from—although the discussion on the material turn in philosophy of technology in some way *compels* us to do so, because of the sheer premises of the debate—but as, simply put, two sides of the same coin. Phrased in our Batesonian vocabulary: a "two" in "one." Karen Barad famously works out her concept of "intra-action" and her theory of "agential realism" on the assumption that matter and discourse are inseparable.[44] In fact, Barad reacts against representationalist forms of realism, that see things—events, phenomena—and words—the representation of those things—as independent realms, but still connected in some way: words correspond to things. That relation is problematic. Agential realism conversely "rejects the notion of a correspondence relation between words and things and offers in its stead a causal explanation of how discursive practices are related to material phenomena."[45] The notion of *intra-action* means to convey—pretty much along the lines of typical relationalist accounts—that "agencies" arise in interaction with each other. There is no preexisting object that then gets represented; there is only "the mutual constitution of entangled agencies."[46] Instead of words and things as separate entities, there are *material-discursive* phenomena (or practices or apparatuses). Like Bateson, Barad finds epistemology and ontology (and ethics) hard to distinguish: "We don't obtain knowledge by standing outside the world; we know because we are *of* the world."[47]

In the field of posthumanism, and in cyborg theory, the indistinguishability of matter and discourse has been a longer-standing insight as well. The cyborg is a construction of matter as well as discourse, of technology as well as imagination, as N. Katherine Hayles elaborately demonstrates—following up on Donna Haraway, who has shown that "cyborgs are simultaneously entities and metaphors, living beings and narrative constructions."[48] Cyborgs aren't just products of the imagination, Hayles points out (then they would merely belong to the realm of science fiction), and they also aren't just "technological" (then they would be only a matter of technical disciplines). "Manifesting itself [the cyborg] as both technological object and discursive formation, it partakes of the power of the imagination as well as of the actuality of technology."[49] And she goes on to stress the importance of *embodiment* in this process—her whole *How We Became Posthuman* is a plea for reinstating embodiment as central even to phenomena such as virtuality. "Embodiment mediates between technology and discourse by creating new experiential frameworks that serve as boundary markers for the creation of corresponding discursive systems."[50] This is a view that comes remarkably close, albeit in a somewhat different sense, to the perspective I advocate here. I see matter (or action) and discourse also as intensely intertwined, but at the same time there is still a remaining *tension* between them. That is the *two* in *one*. We proceed through levels, always taking the view from out of "our circle"—*residing in* matter, but *aimed at* discourse.

There are good reasons, however, for not taking these terms, discourse and matter, too literally—if only because they have been tainted by a whole history of interpretation, and the vague notion of them being two self-standing realms still sticks to them (so deeply ingrained in our intellectual heritage in the form of the distinction between things and words, as Barad also observes, and by extension between objects and subjects). Moreover, "discourse" in my sense—and I would actually rather want to refrain from using the word—does not correspond unilaterally to its meaning as "social imaginary" that is deployed in the humanities and social sciences. What is more, one could surmise, following the dynamic of the central dichotomy, and perhaps readings such as Barad's, that matter equates with relation (fluidity, material agency, assemblages, et cetera) and discourse with substance (reified objects, concepts). But actually this isn't warranted at all. Harman in fact reverses the two, as we've seen. We are "in" substance, in a way, building relational bridges to other "objects."

In a related vein, we might think that societal discourse and social imaginaries still have to wholly catch up with the idea of relationalism, but this also isn't the case—as is masterfully shown by Adam Curtis in his documentary *All Watched Over by Machines of Loving Grace* (2011). Actually an "ecological" ideology of connection, networks, emergence, and self-organization has been defining our time—the idea of networks of computer systems dovetailing extremely well with the neoliberal drive to minimalize (even abolish) state control. Robin Mansell picks up on the same cue in Richard Brautigan's poem from which Curtis's documentary title hails—and which envisions a "cybernetic ecology"—for her powerful analysis of the imaginaries that dominate our thinking about the Internet. And, as I already briefly discussed, she puts Bateson's thought to work in her project. Mansell outlines two social imaginaries that in her view exist in relation to ICT. One can be called neoliberal and is the dominant one. It understands technology as "an emergent and unpredictable process within a complex adaptive system."[51] As a consequence, this process should be left alone; there should not be any intervention in the system, that is, the market system. But there is an alternative social imaginary that could be called cooperative. This model also sees technological change as emergent, but in the context of "a commons-based peer production model."[52] Here too the system should be left to thrive on its own, but from an altogether different perspective—characterized among others by the idea that digital information should circulate freely. Both views, Mansell argues, are linked to two paradoxes: the paradox of information scarcity and the paradox of complexity. The former states, at the same time, that information should be protected by intellectual property rights because it is initially costly to produce *and* that information should be distributed freely because it is almost costless to reproduce. The latter says that the emergent complexity of the system leads at the same time to "loss of control" *and* to "greater control achieved through programming within a decentralized system."[53] Mansell then goes on to examine the conflicts between the two imaginaries—that can be perceived in many societal debates—as illustrations of these paradoxes. Not unsurprisingly she turns to Bateson, as a prime source for navigating paradoxes. The criticism of conservatism that I discussed previously, voiced against Bateson and cybernetics and systems thinking in general, is also referred to by Mansell. But (like me) she believes that he can still be of help:

> His [Bateson's] work has been criticized for its affinity to systems theory and for leaving no scope for human agency or moral values and it has been largely neglected in the social sciences. Nevertheless, it has been applied in the analysis of power relationships within the communication system. [Namely, as Mansell clarifies in a footnote, by Anthony Wilden.] Observations in Bateson's work and that of some of his colleagues can be interpreted in a way that is helpful for understanding the genesis of paradox in a complex adaptive system.[54]

Mansell then proceeds to deploy Bateson's levels of abstraction approach—especially as it gets developed within his theory of double bind, namely, as the result of the confusion of levels—to make sense of the problem at hand. A paradox indeed can vanish as soon as it is made clear that the two "sides" of the paradox belong to two different levels (or, that they belong to the same level, but at different points in time). In view of the tension between the two imaginaries, then, and "with a focus on paradox, claims that policy intervention is harmful can be seen to have a certain validity just as can claims that policy intervention is needed from either above or below. *This is because these perspectives are about the dynamics of change at different levels in a complex adaptive system.*"[55] So, in order to mitigate the tensions between these imaginaries—discourses about technological matters—we need to learn to see them at the appropriate level of abstraction, within the appropriate context. They can exist next to each other (or on top of each other). We don't have to choose between two "stories," for reasons of one being able to represent "reality" better. Mansell's account helps to make clear how discourses are not necessarily a reified, selective business while at the other side of the discourse-matter divide a whole relational world of material networks is waiting to be discovered and charted. Discourses can be relational just as much—at least in content. And from a Harmanist perspective, they *are* relational, ontologically speaking, at any time—by definition. This is to demonstrate that we need to surpass that simplistic image of "one" discourse versus "one" matter, which most of the authors just discussed do not necessarily cherish in that form, but that perhaps because of the attraction and attractivity of "material turns" has to some extent become a commonplace, distorting or preempting the actual view of, indeed, *multiplicity* we need to cultivate in this regard. We need an approach that is able to account for the interrelation of matter and discourse *at any given level, in all contexts*. Yet, this also means reevaluating the central dichotomy in this light—and that means, reevaluating "efficiency."

Efficiency Beyond Discourse

The central dichotomy puts efficiency—more precisely, the value or discourse of efficiency—on one side, wider networks on the other. On one side, we regard technology as just the realization of the value of efficiency. On the other side, we become aware of the fact that technology is *more*, much more than efficiency: it is also the instantiation of a whole constellation of other values, social values. As I've already suggested, in this way philosophy of technology has gained a lot of insights, but it has also to some extent thrown out the baby with the bath water; that is, the baby of efficiency.

Efficiency acquires the status, either implicitly or explicitly, of a something-to-be-surpassed. Putting efficiency on one side and launching calls to attend to the other side makes us surreptitiously turn against efficiency—like a sleight of hand. We lose the capability of talking about efficiency in the wider mode. What is more, this way we also risk leaving "efficiency" in the hands of those who are able to wield the notion as a weapon in an ideological battle. The wider mode of the central dichotomy exactly meant to liberate us, among other things, from ideological blindness. Relegating efficiency to one side leaves, paradoxically, no room for maneuver to question the ideological taken-for-grantedness that nowadays adheres (and has since quite some time adhered) to the term.[56]

Here, too, the history of the field may be implicated in the way efficiency is handled as a concept. Famously, Jacques Ellul, often looked at as a classic philosopher of technology, equates *technique*—translated into English as "technology"—with efficiency.[57] Both are one, in a sense (I will return to this point in the next section). And with technology penetrating all domains of society and life, efficiency becomes the leading imperative (and vice versa). In the Critical Theory tradition, also, efficiency and instrumental reason have long been a target for critique,[58] as one of the ways in which the rational ideal of the Enlightenment gets incorporated and subsequently turns—through a process of "regression"—into a perversion, but a perversion that is societally celebrated and sought after. With Horkheimer: "Efficiency, productivity, and intelligent planning are proclaimed the gods of modern man."[59]

Indeed, Feenberg partly builds on the critical tradition, arguing as well that "the 'natural justice' of our time is efficiency."[60] He goes further, nevertheless, coupling that tradition to the constructivist perspective of science and technology studies. The critical stream, like Ellul, identified technology for a large part with efficiency, but in that way precluded it from being "more"—technology in this view can only be cast aside as a dominating, colonizing force. This is the mirror image of technology being just a means, and nothing "more." Both instrumentalist and determinist stances loom. Technology is in each case seen as the instantiation of efficiency—whether efficiency is regarded as neutral organizing principle *or* as all-encompassing imperative. Constructivism, by contrast, as Feenberg and Bakardjieva assert, "frees technology studies from the dogmatic assumption that efficiency and efficiency alone determine which of the various possible designs of an artifact will end up gaining general acceptance."[61] Feenberg however also goes to some lengths to demonstrate that one of the prime representatives of the Frankfurt School, Herbert Marcuse, can be looked at as a forerunner of current technology studies: Marcuse criticized the "neutrality" of technology and foreshadowed "the notion that technological design is not governed by efficiency or scientific principles but is decided by socially situated actors."[62] At the same time, Feenberg stresses that there is no technology *without* efficiency: "Of all the various ways in which a device can be designed to accomplish a certain purpose, that one will be chosen that satisfies both technical and social criteria."[63] Wha-Chul Son helps to point out that Feenberg shows in essence that efficiency always has to be viewed within a social context. "According to this [Feenberg's] view, the efficiency of existing technologies has been 'constructed' in order to justify their choice."[64] (More on this shortly.) Referencing Carl Mitcham, Son remarks: "the concept of efficiency is context-dependent."[65]

Indeed, this concerns a general characteristic of efficiency—but a tricky one. Such observation, about efficiency's context-dependence, does not prevent that implicitly, au fond, we still cling to a stark distinction between efficiency on the one hand and "its" context on the other hand. This also shows when we look at how the notion of efficiency is wielded in the public debate and in societal discourse. What is efficiency anyway? The way we generally talk about it, cloaks the intricacy of the concept.

In its most skeletal definition, nonetheless, efficiency comes down to a fairly simple idea. Defined in engineering as the ratio of useful output to the total input of a system, generally it serves to describe the extent to which given means are used most optimally in order to achieve a given end. That can apply to just about everything—at least, everything that can be quantified, as there exists a fundamental relation between efficiency and quantifiability. In order to assess efficiency, one needs to *count*. The energy efficiency of one's washing machine—expressed in labels such as A++, A+, A, B, C, et cetera—cannot be accounted for unless it is calculated using relatively precise quantities such as energy expenditure (in kWh), washing temperature (e.g., in °C), and water consumption (e.g., in liters). An efficiency label granted on the basis of vague evaluations like "the machine uses up about a shark's basin of water per year" would be ridiculous, let alone hard to compare with other assessments. But apart from efficiency being by definition a matter of quantifiability, this is to demonstrate that efficiency as such is *goal-neutral*.[66] It doesn't matter to efficiency *itself* that the object under inspection is a washing machine, an automobile, or an enterprise. This is what makes it different from *effectiveness*, a notion with which efficiency is often confused. Effectiveness concerns the extent to which one accomplishes an envisioned purpose. Efficiency is about the extent to which this is done in the least wasteful way with regard to invested resources such as time, energy, money, and so on. In the well-known formula of Peter Drucker: "Efficiency is concerned with doing things right. Effectiveness is doing the right things."[67] Efficiency "stops" before any purpose becomes at stake. But this seems ambivalent. In fact, everywhere we look, efficiency nowadays is embedded in purposive structures. And exactly the wielding of the notion serves to cloak its purposive embeddedness: efficiency in its apparent goal-neutralness succeeds excellently in establishing and consolidating specific societal structures.

Countless critics, certainly not only Frankfurt Schoolers, have delineated how capitalism, and in an even more intense way neoliberalism, is founded on the idea of efficiency. Efficiency is seen here in the first instance in an economic sense: how to *grow* economically, with at least investment as possible? But also the more Foucauldian meaning may be at stake, in the sense of efficiency as principle to steer and guide the process of disciplining. Bentham's panopticon is surely a hallmark of efficiency; due to the prison's construction, it does not even need a guard to make its surveillance system work. Furthermore, there can be a cultural aspect implied, as conceptualized among others in George Ritzer's notion of "McDonaldization."[68] As Carl Boggs comments, quoting Ritzer:

> Workers at McDonalds and other chains are subjected to a matrix of routinized norms, rules, and expectations shaping restaurant operations from one locale to another, designed to ensure uniformity, speed, and predictable results. The outcome is a totally managed system in which "efficiency, predictability, calculability, and control through nonhuman technology can be thought of as the basic components of a rational system"—exactly what Henry Ford might have anticipated many decades earlier.[69]

Some commentators stress the *non-accidentality* of efficiency in the capitalist structure. They point out how efficiency actually *defines* the system; it is not merely a contingent characteristic or coincidental outcome of it. Harald Welzer claims that efficiency, and more precisely the heightening of efficiency—and the command to heighten it—belongs to industrial capitalism as much as labor force and capital. Put even more strongly, capitalism's functioning is conditional upon it; and this has direct ecological consequences. Increasing efficiency must always entail in the end a further depletion of the earth's resources, Welzer argues. Thus, capitalism is essentially a predatory system in relation to the earth's ecosystem. "Increasing efficiency, which can never mean anything other than the efficiency of raw materials, is part of its essence [i.e., of capitalism]."[70]

Such analyses would seem to "essentialize" efficiency all over again: localize it within the foundation of a political-economic system, and marking it thus as something to be overcome. The situation, however, is more complex. Efficiency is in practice not purely goal-neutral (instrumentalist view), but it's also not all-pervasive (determinist view). This becomes clear in an analysis such as Janice Gross Stein's, which zooms in on the "cult of efficiency." Efficiency, Stein demonstrates, has become an "end in itself," while it should be "a means to achieve valued ends."[71] When this happens, when we mistake the means for an end, we get a cult.[72] Political interests instigate this cult: "The cult of efficiency, like other cults, advances political purposes and agendas. In our post-industrial age, efficiency is often a code word for an attack on the sclerotic, unresponsive, and anachronistic state, the detritus of the industrial age that fits poorly with our times."[73] Efficiency, as discourse, means to consolidate (neo)liberal market thinking.

It is in this sense—*as cult*—that the principle of efficiency gradually, eventually, and not much unlike Ellul observed and foresaw, tends to seep through to domains that were once thought to be "efficiency-free," or at least freed from its absolutist dictate. Health care, education, academic research, and the arts spring to mind. In all these domains, an efficiency imperative is sounding louder and louder by the day. Obviously, global economic and political conditions—financial and economic crises and, mostly, the attendant political "austerity" measures—have a significant role in this. Nonetheless, the overall effect is that an "efficiency discourse" gets chiseled more and more into the hearts and minds of people. Efficiency, as the idea and *ideal* of optimizing the use of means on a quantifiable basis toward the realization of a pre-given goal, becomes a natural thing to aim for. This has as a logical, apparently trivial, but actually pertinent consequence that any call for or even tolerance of *in*efficiency—perhaps for reasons of the simple observation that some things cannot be quantified in order to compare them—is instantly cast aside as *un*natural.

This makes the sober scrutiny of efficiency as a concept, in a societal way—that is, in the societal debate—awfully hard. We can easily give a simple dictionary definition of it, but we can hardly accurately grasp how it works, that is, in all its societal dynamics, as the notion got so integrated into our everyday thinking and talking. This way, we could be easily seduced into casting it to one or other side, as in the case of the central dichotomy, and *forget* it in a certain fashion.

The approach that I have developed throughout this book offers the advantage that with it, we can *find back* efficiency, by "relocalizing" it, or better, *topologizing* it. With levels of abstraction and the remodeled discourse-matter dichotomy in mind, we can ask: at what level of abstraction, within which context, are we looking at efficiency—and are we envisioning it from its matter or discourse side? Once more, the terms "discourse" and "matter" should not be taken too literally, or understood in their classic meanings; rather, it is about "what we do" versus "what we say," much in the Latourian sense of proliferation versus purification. Indeed, historical investigations show that efficiency hasn't always been what we now take it to be—and shouldn't even nowadays be understood linearly in one sense. Jennifer Karns Alexander in her impressive historical study *The Mantra of Efficiency* traces the efficiency notion down its genealogical trajectory,[74] showing how the term before the nineteenth century could be found mostly in medieval scholasticism, signifying a philosophical concept that meant to convey the superiority and wisdom of God. The roots of that use went back to Aristotle's efficient cause (*causa efficiens*). Stein, also, observes how "the ancient Greeks spoke of efficiency, but as a means, not an end, of politics and society" and how "in Platonic thinking, the purpose of efficiency is virtue."[75] Efficiency in classic thinking is, quite literally, a matter of the/a greater good—for, as we saw in Berman's analysis as well, this kind of thinking starts from a wider-purposive structure, in which all things are seen to be embedded. One must imagine the term "efficiency" to have a totally different flavor in such an outlook than we are currently used to (and have been used to for some time).

Thus, for medieval theologians, picking up on Aristotle, God is the first efficient cause, or moving cause, of everything that exists, and God manages the cosmic household, one could say, as efficiently as possible, wasting nothing: every creature and thing in his creation take part in his divinity. Another source to which Alexander points is Ockham's razor, the famous formula of William of Ockham that says that *entia non sunt praeter necessitatem multiplicanda*: entities must not be multiplied beyond necessity. In finding explanations for phenomena, one should pick the hypothesis that presupposes the least entities. But again, this is mainly a philosophical, that is, epistemological, matter. It is only starting in the nineteenth century, with the Industrial Revolution, that the term "efficiency" becomes applied to human beings and human organization structures. Very soon, an "efficiency craze" would sweep Europe and the United States. Today the best-known proponent of that culture is probably Frederick Taylor, the founder of "scientific management," who set out to make labor processes in factories more efficient by mapping them and reducing them to their smallest units, so that he could determine the least wasteful sequence of movements, actions, and decisions.

It is interesting and remarkable that "our" very own efficiency craze—storming among others academia, health care, and education—has a forerunner of about a century old. Instead of expecting miracles of efficiency from industrial technology, however, we nowadays put our hopes in ICTs. As Evgeny Morozov sarcastically remarks, "As long as information is produced and processed efficiently, the legacy of the Enlightenment is believed to be in good hands."[76] But crucially, the craze—the ideal, the ideology, the discursive dictate stating "this is how it should be; this is what you should strive for"—is not identical to the material, that is, technological, societal constellations in which efficiency is cast; in other words, the map is not the territory. And apart from that, one kind of efficiency does not have to equal another. A way to visualize this, I propose, is to talk of *levels* or *places* of efficiency—hence, topologizing it. Efficiency can be seen to unfold at different levels. There are different kinds of efficiency, but we need to distinguish between them.

We may consider, to begin with, something like biological efficiency. While the engineering definition of efficiency was seeping into societal debates throughout the nineteenth century, it also affected the sciences. Charles Darwin, for one, as Alexander elaborates, gives the notion an important role in his work. The correlation, for instance, between the size of a certain genus and its success in adapting, is a sign of a more "efficient" mutation. Stein finds the concept back in nineteenth-century studies of "human energy," and remarks how "the language and the imagery of these early physiological studies of human energy are striking. . . . We are inherently, innately, superbly efficient. Efficient is what we are."[77] Seen from a certain perspective, living beings are indeed strikingly efficient, as species, organisms, but also as individuals, though always in a (social) context. Babies and toddlers develop in a remarkably spontaneous way a multitude of skills in an extremely short amount of time: grasping things, use of language, cognitive skills, and so on. As humans we do take a long time, in contrast to many other animals, to reach a stage in which an individual is able to survive on its own, but the ingrained efficiency, however relatively slowly deployed throughout a social-educational context, is still impressive.

On the level of social groups and culture, then, we may think of for example cooking skills and rituals that have evolved throughout centuries, handed down over generations and fine-tuned and perfected through practice. Those kinds of practices, too, are a matter of efficiency. We may not remember this in our affluent twenty-first-century Western world, but for long periods in human history and in many places, people had to make do in situations of scarcity. Cooking in such contexts becomes preeminently a matter of wasting as little as possible (with perhaps the usual, recurrent rituals of extravagance and wastefulness in between, during feasts or holidays or on the occasion of the visit of guests).

From another perspective, on a still higher level, then, we have on the "societal" level the wider efficient functioning of the economic system that I already touched upon when discussing efficiency as a core value of capitalism. Looking at it even more broadly, one may see an efficiency mechanism deploying itself throughout the history of societal constellations. This is in any case the point that Daniel Bell makes in his wide-ranging analysis of *The Coming of Post-Industrial Society*.[78] Bell outlines three types of society that follow upon each other in chronological order: agrarian, industrial, and post-industrial. Now, the "switching mechanism" between them, what makes one type transform into the next, is technological development. From a certain moment in agrarian society, technological innovations start to make it possible to produce food on a larger and more efficient scale; this means that, gradually, many people who previously had to be occupied with growing their own food become available for industrial labor. Something similar then happens in the progression from industrial to post-industrial society: because of automation and optimization of production processes, labor force becomes available that can again be put to use elsewhere, namely, in services. Webster calls this "mechanism" in between, grounded in technological innovation, the " 'more for less' principle"[79]—which is as elementary a definition of technology (and efficiency) as one can get.

I already spoke of "everyday efficiency"—the efficiency of guitar playing, nourishing an intimate relationship, raising children—and this can be seen to crisscross some of these levels. Efficiently tending our garden, for one, is a matter of biology, culture, and individual (and social) practice. And we can find so many places in which efficiency can be found back. There *is* something like a societal "efficiency ideology"—one can feel it in many domains of life. But there is no *one* discourse on efficiency, as there is no *one* realm of action in which it is realized. At each level, in each spot, in each place, we may relate from out of *that* place to a certain ideal, an aim that is discursively placed before us, to which we need to aspire. From this perspective, it makes sense to speak of many different kinds of efficiency, that may or may not be in conflict with each other. According to some, for instance, ecological efficiency is not at odds with economic efficiency, while others (Welzer is one of them) find them to be completely diametrical. But the fact of the matter is: just like we found a multiplicity of discourse-matter relations, we have found an abundant wellspring of efficiency beyond the narrow mode of the central dichotomy, beyond "discourse."

Manifolding Purpose, Reclaiming Efficiency

With the multiplication of efficiency, its unfolding along a colorful spectrum of varieties, comes the manifolding of purpose. Once again—I cannot repeat it enough—this is not to plead for the all-reigning dominance of efficiency-as-we-know-it. There are narrow efficiencies: discursive ideals formulated as a dictate: "do x or y, because it is efficient." In such phrases the term "efficient"—often vaguely defined but, paradoxically, perfectly understood by everyone—serves as final legitimation for a furthermore undiscussed societal, organizational, or existential measure, act, or rule. This is indeed efficiency as we know it: instrumental purposiveness. And in this meaning, efficiency has obviously gotten something of a bad rap among those who purport to see through the ideological veil of deception. Efficiency poses as neutral, as commonsensically natural and logical. "Who could possibly be against the concept of efficiency?"[80] But the term and the value of efficiency, in their powerful taken-for-grantedness, are, as Stein points out, "invoked to escape politics, to depoliticize the provision of basic public goods."[81]

To be sure, this situation still demands that we denounce or, better, deconstruct efficiency to the extent that it is regarded monomaniacally as holy aspiration. This makes a framework such as Ellul's still highly relevant, and supremely prescient moreover. Ellul looked around in society when he was writing books like *La technique* (in the 1940s and 1950s) and sought to outline the dynamics in what he perceived; but rather than having become outdated, his insights seemed to have gained in pertinence. As Alexander observes, "Enthusiasm for efficiency remains high; only in intellectual circles have critiques such as those by Ellul and Mumford been seriously engaged."[82] Nonetheless, we've seen how contemporary philosophers of technology generally tend to eschew Ellul, under the consideration that his analysis is determinist or essentialist. This might be a too-quick evaluation. Wha-Chul Son in the aforementioned paper and an earlier one works out a helpful way of interpreting Ellul in a nondeterminist manner, but without sacrificing the gist of his argument.[83] Son defends Ellul's analysis of the "efficiency principle," but on the condition that it is "understood as describing a situation in which everything is justified *in the name of* efficiency."[84] This means a slight correction to the outright *equating* of technology with efficiency. Such maneuver immediately preempts a possible misunderstanding of Ellul as saying that all technology is *efficient*. Clearly, sometimes technologies also heighten *in*efficiency. "Technological development is not pursued in order to achieve more efficiency, but *in the name of* more efficiency."[85] That is, efficiency, much along the lines of our above discussion, serves to legitimize technological organization, and hence a certain form of societal organization. "All techniques are supposed to be, and justified as, the pursuit of efficiency."[86] Ellul's efficiency principle, thus, in Son's reading, refers to "a situation in which people accept any device or activity once it has been qualified as 'efficient.' "[87] So it is actually not *really* about efficiency, understood as the search for optimal input-output ratios. Things might not *actually* be efficient: "Different gadgets and activities are explicated as efficient without clear evidence or support. People are ready to believe it, not bothering to calculate the actually [*sic*] efficiency, even when it is possible."[88]

In my view, however, this is a good reason to *expand* the notion of efficiency, rather than have it secluded within some safe realm in which it is reduced to its supposedly real contours—namely, in philosophy of technology's terms, to one side, that is, the narrow side, of the central dichotomy. As we've seen, efficiency is omnipresent (like breakdown) and we can develop a much "warmer," relational, ecological, wider understanding of it. Why? Because this *really* enables us to take heed of wider purpose(s). Instead of cautioning against efficiency's ideology-consolidating dynamic, waving our hands in a dismissive gesture, saying, "It's *just* efficiency, people, nothing to see here, move along quickly!" we should say "Efficiency is *much more* than that," much more than merely an instrumentalist *or* a determinist interpretation can make sense of. Efficiency in that broader sense can and should actually be a wholly righteous guideline for existence, a creed for life—though certainly not in its narrow form. When hailed in its narrow form, on a cultural-societal level, efficiency becomes the harmful imperative that we know so well—a cult. But when cultivated and cherished in its wider, broader form—especially also in the light of the notion's historical origins—it may become a wholesome part of an art of living with technology, making it truly possible to engage with purpose(s).

Indeed, commentators have offered the notion of purpose to complement or correct for efficiency's one-sidedness, or more precisely, its ideological functioning. Stein remarks upon the relation between efficiency and effectiveness, observing how it is often presented as if the two are mutually exclusive, that one has to choose between them, while in fact "effectiveness is built into any concept of efficiency."[89] So, efficiency is not *truly* goal-neutral; there are always goals at stake, albeit that talk about efficiency *within the cult* cannot make sense of those. Stepping out of the cult and moving to "analysis," in Stein's view, can only be done by making explicit and talking about the goals. What is the purpose that this or that efficiency is meant to fulfill or support? "What distinguishes the legitimate conversation from the cult is a serious and deep discussion of purpose: at what do we want to be effective?"[90] And there is no doubt that this is a "political conversation."[91] Son, too, places a renewed engagement with, or an explicit discussion about, purpose in opposition to the efficiency principle. The efficiency principle makes goals "become redundant," he observes. Efficiency itself, not the goals it is meant to achieve, becomes the driver for technological development. With a reference to Langdon Winner's notion of "reverse adaptation," Son remarks: "When a goal for a technological innovation is suggested, it is either faked or redefined in terms of efficiency so that the current path of autonomous technology continues.[92] Winner . . . discusses the latter case calling it 'reverse adaptation' referring to the case in which the end is readjusted to fit the means."[93] The core problem is not efficiency as such; it is the fact that goals, and the contexts in which efficiency is to be pursued, are not clearly defined. So quite in line with my argument here, Son asks: "If the final goal of a practice or an artifact is good, then why not try to find an efficient way to realize it?" He gives the example of an environmentally friendly car: looking for the most efficient way of manufacturing such a car is a good thing to do. One could add to this nevertheless that drawing the frame too exclusively in terms of an artifact "car" runs the risk of obliterating wider network concerns—questions such as: should we strive toward developing cars as our most appropriate means of transportation, or are there better, albeit more "radical," alternatives, such as other types of mass transportation? To that extent, *without* consideration of these concerns, just efficiently producing that environmentally friendly car—the effect of which may be that other options are blocked out or soft-pedaled—might *not* be such a good thing. In any case, Son then goes on to elaborate the notion of "purpose driven technology" (the term "purpose driven" derived, but without further conceptual connection, from American pastor and author Rick Warren's book titles *The Purpose Driven Life* and *The Purpose Driven Church*). "The development of a technology . . . should begin with a justification of its purpose."[94] Efficiency concerns *can* and still *need* to play a role, of course, but in their right place: "The criterion of efficiency, in a space-and-time-specific sense, could be used for minor and concrete decisions but not for the justification of a given project."[95]

Now there are two potential problems with this. Opposing, or better, complementing efficiency with purpose is one thing. A more pragmatically oriented observer might comment that in all technological design, purposes are of course considered. Ask Mark Zuckerberg why he develops Facebook and you will probably get as answer an array of purposive legitimizations: to connect people, to strengthen the social fabric, to make a better world, and so on. Or, with another example, heightening efficiency in an academic setting—by way of the installation of a whole infrastructure of follow-up procedures, target achievement trajectories, "points" collection systems, et cetera—might easily be explained in terms of purposes such as: to make it a better organization, to enhance quality, to demonstrate our societal relevance. Purposes abound; there's really no lack of them. Second, it will not necessarily do to offer "a" purpose or even a set of "purposes" as an antidote to a purportedly one-sided courtship with efficiency. Son suggests, in terms compatible with my investigation here, "The purpose of technology could be considered on different levels, from general to specific. One could search for a general purpose of technological innovation as a whole, while others focus on a particular technology."[96] Indeed, such a multilevel approach would be laudable; however, we would have to make sure not to fall in the trap suggested above in relation to the example of the environmentally friendly car: emphasizing a purpose at one level might be done at the detriment of (a) purpose at another level, whether broader or more specific (but most of the times probably broader; as Bateson knew so well, we're still inclined to mostly neglect broader systemic factors).

In response to these potential issues, we must, as I already suggested, stress the *multiplicity* of purpose *as well as* of efficiency—to the extent that it would be a mistake to present efficiency as "one" and purpose as "multiple." Doing the latter would just involve, once more, the relegating of efficiency to one side—of the central dichotomy—and thus to be done with it. But the first issue mentioned above, namely, that there is mostly no lack of purposes at hand to explain and justify certain actions, decisions, or technological developments, exactly shows how in all talk about technologies efficiency and purpose are inextricably intertwined. It is our job to unravel these knots, map the values involved, in order to really evaluate where one value ends (and should end) and where another begins (and should begin). If we cast out efficiency too quickly along the lines of an adage such as "It's just efficiency, there are more important things at stake," we will lose the clout to oppose claims such as "But what could be wrong with efficiency?" Phrased differently, efficiency is too commonsensically important to have it stepmotherly treated as "just this, and nothing else." We don't need to counteract it, in the pure sense of the word: we need to *reclaim* it. Concomitantly with this, purpose(s) should be investigated at as many levels as not only practically but also epistemologically possible; which links up to the second issue. Another way of putting this is that we need to align anew the "talk" and the "walk," in the spirit of talking the talk versus walking the walk. That is, we need to not just symbolically talk about efficiency, oppose something else to it at the discursive level (see Puech's critique of remaining within discourse). That would only result in staying at the same level as the efficiency discourse, where, as Johnson remarks, "often the primary purpose of participating in innovation is symbolic—a demonstration that one is forward looking and modern, willing to jump on whatever bandwagon may be rolling by."[97] No, we also need to *live* efficiency[98]—which is, to be clear, something altogether different from living *efficiently*—but from a multipurposive perspective. Of course, the framework that we've constructed on the basis of Bateson's thought is ideally suited for this—which can be illustrated once more, by taking a last look at Bateson's cultural critique.[99]

Learning from Bateson's Cultural Critique, to Become Artful

The art of living with technology is in the first instance a human endeavor, no doubt. It is an issue for "us" existentially speaking, and ethically speaking. As is nowadays increasingly investigated by some scholars, (other) animals might "have" more technology than we long assumed.[100] Do animals also develop something like an art of living with/through/in technology? Perhaps they do, although we might find it difficult to demonstrate this. But if an art of living with technology along Batesonian lines entails the navigation across contexts, they may share in this art to an extent—at least as much as they share in the "pattern which connects." This is the paradoxical situation we are in—although it is not really paradoxical; all depends on the level of abstraction at which we look at the matter—of a characteristically human endeavor such as the artful navigation of purposive structures turning out to be essentially a question of moving beyond the merely human perspective. With Harman, we've seen how stretchable even that expandability is. It is not just a matter of taking into account other living beings (animals, plants, cellular organisms). In times of quasi-autonomy-acquiring algorithms and big data structures taking on "a life of their own," we can deploy Harman's object-oriented perspective,[101] if only as a tool for imaginary exploration, to chart the contexts-within-contexts that are wrapped up in the purposive constellations that we, ideally, should want to extensively map in order to *decide*, in a manner as aware and as informed as possible, about what to do with technology—which is: what to do with ourselves, with our lives, society, the world. This means, however, no longer puzzling over *technology*—we've lost, or at least are losing technology—but asking about *purpose* in the first instance.

We do not need to plunge headlong into that exercise shouting: "Purpose, purpose! Where is purpose?" There is a systematicity that we can bring to the table, and this on the basis of the framework built here around the skeletal structure of the central dichotomy reframed from a Batesonian angle. Fundamentally, we've multiplied, manifolded the central dichotomy—endlessly. The central dichotomy characterizes all contexts, given that from out of each context we look out onto another one. "We" here can be meant phenomenologically: we as observers, who from a lived perspective experience the world; but it could also be a generic "we" (rather imagined or hypothesized by us) as a way of taking the standpoint of an other, which could be any object in the Harmanist sense: a species, an organization, a society, a business plan, an artifact. A first question to ask, then, would be: are we in a given context chasing after a conscious purpose—that is, unilinearly expecting a result x from an action or set of actions y—and to that extent *not* taking part in an awareness of the given that we *are* in this context, thus, unable to see the wider context? Or are there broader purposes, other purposes, we are losing sight of? Are we chopping off the loop structure, or not? Is our view "half" and in that sense distorted, or do we attempt to acquire a wider two-in-one systemic-ecological awareness? The fact that this kind of thinking steers us potentially in the direction of an infinity of possibilities should not intimidate us. We are aware of that infinity, and have no pretense about grasping it in whole (following Harman, no thing is wholly graspable as such). We just try to locate, specify, and describe our position as accurately as possible; no matter where we are, in actuality or in our imagination.

There are patterns, Bateson would remind us. The point is that we have to (re)learn to see thing structures as purposive structures. This was the revolutionary insight offered by early cybernetics, which started to use the notion of purpose—reinstating "teleology"—to think about the behavior and functioning of systems. Such a move must have come as a shocker after centuries of trying to frame nonhuman nature as a dumb reservoir of building blocks and resources. The problem is that in the meantime we haven't *really* unlearned the Cartesian worldview, bequeathed to us through the generations. The early insight from cybernetics did not really seep through to our collective consciousness. But we did receive and put to ample use cybernetics' practical outcome, that is, computer science, in tandem with computer technology. So, in a way, we got the power, but without the wisdom.[102] We got the conscious-purposive enhancements, without the multipurposive ecological awareness.

Obviously, purpose as we generally know it is a highly personal matter, fraught with existential uneasiness. As Berman already commented, from modernity onward, we have been challenged to find our "own purposes" instead of having them prescribed or handed down by religious or political ideologies. We are stuck in between, on the one hand, instrumental purposiveness, instantiated in our tools, technologies, and sciences, which help us get a grip on the world and on nature, and on the other hand a vaguer, more elusive existential purposiveness, to be filled in and colored in by us, possibly with the help of (some of those old) religious, spiritual, or other worldviews (or possibly gurus). Why not simply connect these two? It is bridging the central dichotomy, but from an existential standpoint. Let's *reclaim wider purpose for technology*, and *reclaim narrow purpose* (instrumental rationality, instantiated in technology) *to make sense of life*.

How to go about thinking through purpose—in multiple senses of the word, that is, "considering carefully" as well as "thinking via"? We don't have to take up the task *in vacuo*. We are not completely dropped in the wild. As Bateson says, there are patterns, and the dynamics of how these patterns arise can be deployed as tools that help us cope with *purposive discussions and decisions*. I have been building up the general framework throughout this book, but I want to focus once more, as a way of rounding out the argument, on a couple of aspects that pertain especially to Bateson's cultural critique and to that extent have direct, tangible ethical consequences.

Contexts come about within other contexts, according to the two-sided ontological scheme that Bateson outlines. Contexts are nested into each other. However, depending on which side of the scheme we choose to emphasize, distortions will begin to occur. We've seen how Bateson cautions for a one-sided emphasis on quantity and on the digital. This critique harbors some ambivalence, as we've also discussed, but the categories are still useful and can be made workable even—actually all the more so—in relation to our contemporary situation.

We've discussed the "materialist superstition" that Bateson diagnoses: the belief that quantity determines pattern, which is, he adds, "a basic premise in contemporary economics."[103] It amounts to an explanation in terms of one half of the full picture, according to which matter (Pleroma/flux/quantity) and mind (Creatura/structure/pattern) are interrelated, interwoven, bridged—but still essentially apart as well. Patterns come about as marks into the flux of matter, and none can be explained in terms of the other one (as said, there is also an "antimaterialist superstition," cherished by some spiritualisms, new age beliefs, and so on). Remember the metaphor of the chain that breaks under tension: the exact place at which the chain breaks cannot be predicted—it is a wholly new event, a pattern that comes forth, of and by itself, however from and within the stream already present. According to Bateson, culturally—that is, in the Western world, and in our current times—we have the greatest difficulty in taking this into account: "We are very impatient, you see, of the idea that there are patterns, that the patterns have rigidities and rigors which have to be respected."[104] Patterns do what they do; they emerge and then they are there, and we really have no choice other than accepting them, accepting their "tyranny," the "enormous tyranny of patterns,"[105] instead of assuming we can unilaterally steer them, more precisely, by controlling and moving around quantities. Typically, that is what governments do, Bateson suggests—although the idea could be called a culturewide presupposition by now:

> Consider the business of government. The government has control over *quantities*. They can alter the tax rates and they can manipulate quantities in various ways, but the trouble for government is that the weakest link is never predictable. You can impose quantitative change upon the system but you can never tell what the outcome will be.[106]

This, of course, relates to the deceit of conscious purpose—instigating in us the assumption that we can *solve problems*. We never simply can, no matter how much quantification or quantitative analysis we apply. Basing our whole outlook and societal organization on this principle and on these presuppositions leads to disaster. Bateson: "The way we are going about things with this enormous emphasis upon the quantitative view and the minimal emphasis upon the patterned view is, I believe, the easiest way of descent into hell. The surest."[107] He mentions the use of grades and tests as an example of this way of thinking, of "quantification imposed upon a patterned, biological world."[108] How often do we not approach things in those terms? We probably do this to the extent that we can hardly distinguish anymore, within our discourse, how dominated by quantitative reasoning our ways of thinking and talking are (Mary Catherine Bateson asks, "Have we Pleromatized language?").[109] Obviously we can spot here a direct overlap with the efficiency discourse. Efficiency, as was pointed out, must be quantified in order to be measured. And there is a fundamental link, historically, between efficiency's purported neutrality and quantification's success. Berman grasps and summarizes this beautifully in this longer quote:

> The same class that came to power through the new economy, that glorified the effort of the individual, and that began to see in financial calculation a way of comprehending the entire cosmos, came to regard quantification as the key to personal success because quantification alone was thought to enable mastery over nature by a rational understanding of its laws. Both money and scientific intellect (especially in its Cartesian identification with mathematics) have a purely formal, and thus "neutral" aspect. They have no tangible content, but can be bent to any purpose.[110]

This is, naturally, the context in which, as a reaction to the dominance of quantity, calls are launched for a rehabilitation of the concern for *quality*, and for effectiveness—the latter being characterized by a "focus on quality rather than quantity."[111] Such rehabilitation implies indeed a rethinking and reconsideration of "non-neutral" purpose. However, importantly, this shouldn't come down to an overcompensating maneuver: placing all our bets on quality *instead of* quantity. In a Batesonian vein, again, we must cultivate the *two*-sided view.[112] Stein puts it well when she observes in the context of a discussion on efficiency and accountability in health care that "if quality of care is the goal of the health-care system but quality is neither measured nor evaluated, accountability, just like efficiency, remains at the level of rhetoric."[113] As said, we *need* efficiency—redefined in the way I propose—and so we need quantification (and abstraction as well, because, as we also saw, efficiency demands abstraction). The crux of the matter is: we have to strike the balance between the two, an exercise that is, admittedly—as so many of us can attest to from daily life and work in organizations, companies, government institutions, universities, and so on—excruciatingly hard, but nevertheless the only way. It is either "completing distinctions," with the phrase of Douglas Flemons—that is, on a higher level of abstraction becoming aware of the dynamics of how a certain distinction works its effects, in this case the dichotomy between quantity and quality—or choosing between the two poles of the distinction, that is, simply remaining "within" the bounds of the context that is shaping our whole way of looking, merely working with the terms provided by it, and staying oblivious of the idea that it's the either-or choice *itself*, namely, on the lower level, that is hindering us.

This then brings me to the next point. Convergent with the tension between quantity and quality, as we've seen, is the digital-analog dichotomy. These two dichotomies seem to largely overlap at first sight, yet at the same time they do not, because, as said, there is an inconsistency in Bateson's cultural-critical evaluation of the two sides: he defends pattern against quantity, but then he turns out to be also defending the analog against the digital. So which side is he on? The ambiguity remains unsolved. Yet it would appear that beyond this ambiguity, for Bateson, the central matter is the preservation of a "full" epistemological perspective (and conversely, the danger of distortion thereof). Like quantity and quality, the digital and the analog make for two completely different, even incommensurable perspectives. The digital (structure) makes marks in the analog stream (flux). As Berman puts it, "Digital knowledge makes itself evident by 'punctuating' analogue knowledge."[114] And he observes how Bateson came to the realization that "analogue and digital modes of knowing were not really mutually translatable."[115] But historically we began to prefer the digital over the analog. "In premodern culture, the digital (when it did exist) was the instrument of the analogue. After the Scientific Revolution, the analogue became the instrument of the digital, or was suppressed by the latter entirely, to the extent that such suppression was possible."[116]

Now, what is so wrong with the so-called distortion that the digital brings? The issue is that the digital works with the binary pair of 0 and 1: an either-or choice. It's *either* this *or* that. When this kind of thinking becomes dominant, it starts to shut out *both-and* situations. More precisely, those kinds of vaguer, richer situations are simplified, narrowed down to a "yes or no" frame. Anthony Wilden elaborates this, hinting at how pervasive and impactful digital modes of communication and thinking can become (also talking about the iconic mode of communication, that I leave undiscussed here):

> In modern society . . . where emotion and feeling, which are typically analog or iconic states, are commonly disparaged and divorced from (digital) reason (in spite of their being its ultimate source, i.e. the reasons for reason), and where images and thoughts not expressed or expressible in words are commonly regarded as not worth thinking at all, digital information and digital coding—the form taken by signification in language, by commodities in production, by money in relations of price, and by value in relations of exchange—have come to dominate their context of analog and iconic communication, just as "all-or-none" has come to dominate "more-or-less," just as "either-or" has come to dominate "both-and," and just as competition has come to dominate co-operation.[117]

Now one may ask: is such a gloomy depiction of our epistemological outlook warranted? Aren't these concerns about distortion exaggerated? Well, look at your average smartphone app. Apps are digital technologies. That means, of course, their underlying structure is "digital"—digital computer technology. But take some time to become aware of how many apps also work on a "content" level with *digital categories*—albeit not literally in the sense of binary code, but certainly in the sense that they deploy a framework of either-or categories and choices. These choices and categories then have a further impact on how we think and act in the world. One striking example is the Belgian Realo app that stirred some debate a couple of years ago[118] (similar services had been developed before in the United States). Realo is a real estate app through which you can browse for houses for sale or for rent. Nothing too special about that, were it not for the fact that you can also consult the "social profile" of a certain neighborhood: you can look up your potential future neighbors' average income, their age, how many singles there are in the neighborhood, and how "multicultural" the area is. The app is not doing anything illegal; all data is derived from publicly available databases. The application just aggregates this data and presents it in a handy interface. Nonetheless, important ethical questions can be (and have been) asked about possible discrimination. But even more fundamentally, there is the effect of casting a crucial life decision and existential activity, the buying or renting of a property, in (possibly literally) black-or-white frames. Intuition, a critical faculty in making such decisions—how does this house or this neighborhood *feel* to me, does it *feel right* for me?—is potentially short-circuited in this way. The same concern counts for many other apps. The subjective stream of experience—typically a more-or-less or both-and phenomenon—becomes grasped into, reduced to rigid, perceivable quantities or sets of data. A neighborhood becomes a map with quantities; a person becomes a profile; experience or memory a set of photographs on a timeline with tags and links.

Undoubtedly these apps, and algorithms in general, can also *surprise* us. This is the thrill, the exhilaration of big data, of aggregated data sets. Suddenly all sorts of new shapes and patterns pop up that were inaccessible up until now. But the question is: where does one start? If one starts at the digital and blocks out the analog beforehand, the two-in-one dynamic is lost. This has huge implications for how we relate to each other, to the world—and how we make our world. Mansell comments, much along the lines of Berman and Wilden: "When the digital process of communication starts to be dominant in human relationships, this can be seen as a 'mispunctuation' of the communication relationship into binary oppositions which are presented as either/or alternatives, when in fact they are both/and relationships."[119] In fact this critique corresponds with the critique of philosophy of technology's overly empirical bend. Empirical methodology also requires—we saw how Smith has made some remarks to this extent[120]—that one isolates a phenomenon in such a way that it becomes clearly perceivable, calculable, datafiable. In and of itself, in the case of some phenomena—such as subjective experience—this can only lead to a diminishment of their richness and breadth. According to Berman, we can only counteract this by retrieving "archaic modes of thought," however, without reformulating these "in empirical-conscious terms"—that is, in our contemporary scientistic vocabulary. Because, as Bateson suggests in Berman's view, this is "to destroy them [the archaic modes of thought] in the name of understanding them."[121] ("Understanding" obviously is meant here in the sense of *attempting* to understand, but really in the process deforming one's subject matter.) For Mansell as well, who argues in accord with Bateson, there is a way out of "mispunctuation," in the form of "adaptive action," "such that a circular, self-referential, communication circuit can be disturbed by metacommunication," where metacommunication is understood as "communication about communication."[122]

In sum, as with quantity and quality, we need both modes: analog and digital. The both-and of analog heaves us out of the constriction of either-or constellations. Yet the digital is there, too, it is inescapable—either-or choices are needed to push things forward: shall I take this train or not? Shall I use this technology or not? Shall we develop this technology or not? And so, we are back at purpose. "Quality-quantity" and "analog-digital," as conceptual tools, help us to think—at first sight apparently in an abstract way, but actually in the end very concretely—about purpose and about levels. A Batesonian art of living with/in/through technology basically concerns the navigation in and through levels of abstraction. This is, as we saw, what art is in essence about, according to Bateson: art is concerned with the relation between levels of abstraction. "Artistic skill [meant here in the more standard, i.e., narrower sense] is the combining of many levels of mind—unconscious, conscious, and external—to make a statement of their combination."[123] "External" here is to be understood as the structures of "mind" surrounding us everywhere: the ecology of mind of which we are (just) a part. Nonetheless, art does have a special relationship to conscious purpose, as we also saw, and to that extent to (possibly unconscious) multipurposiveness. Art is concerned with "correcting a too purposive view of life and making the view more systemic."[124] This once again does not entail a blunt reversal to the unconscious—notwithstanding that Bateson places "the unconscious" in opposition to "the deliberate."[125] Becoming artful, one might say, is a matter of turning *inward*, but at the same time, that maneuver constitutes a move *outward*, of recognizing "another mind within our own external mind."[126] One watches the flower, realizing how it "works" in the same way as all living things. "The 'primrose by the river's brim' is beautiful because we are aware that the combination of differences which constitutes its appearance could only be achieved by information processing, *i.e.*, by *thought*."[127] That is: thought beyond conscious purpose.

In being oriented too narrow-mindedly toward conscious purpose—"the appetitive"—we've lost touch with "the aesthetic." We've lost touch with what connects us humans to the rest of nature, Bateson suggests, for "ultimate unity is *aesthetic*."[128] As Mary Catherine Bateson comments, "All organisms—not just art critics and philosophers—rely on aesthetics all the time."[129] Our "loss of the sense of aesthetic unity," Bateson presumes, "was . . . an epistemological mistake."[130] Of course, what I have done here is to try to expand Bateson's framework even further, by way of Harman's OOO. Could this widening of the realm of the aesthetic apply to all objects in the Harmanist sense? Harman himself has actually hinted at such a possibility—although up until now without really systematically elaborating the suggestion, and with some remaining ambiguity. *Art* in Harman's view is one of the many possible ways in which fourfold objects can interact. More precisely, it is an instance of what he calls *allure*, which he defines as a "fusion" between on one hand the sensual, changing qualities that we can perceive of a thing— for example, as illustrated by Husserl, the shifting profiles that we see of a house when we walk around it—and on the other hand the withdrawn real object "behind them." On the basis of perceived sensual qualities, the real object can only be hinted at: through allusion. "Instead of the direct sort of contact that we have with sensual objects, there is an allusion to the silent object in the depths that becomes vaguely fused with its legion of sensual qualities."[131] And this is the domain of art (art is one type of allure; not all kinds of allure are art).[132] To put it somewhat disrespectfully: it is art's *core business* to hint at real objects through allusion.

Still, there is a remarkable aspect to Harman's reflections here. His OOO has as its general premise that it counts for *all* things. But in a couple of lectures, Harman has specifically mentioned that in his view art is something pertaining to humans alone.[133] Surely, that is a thesis that is self-evident and convincing enough. But one is left to wonder, certainly with Bateson in mind: should we really constrict art or, more generically put, the *artful*, to the human domain? Harman loosens the conceptual tension by clearly granting more scope to the notion of the *aesthetic*. In an interview with Erik Bryngelsson, the latter asks whether the Harmanist framework could be described as "a kind of 'aesthetics as first philosophy,'" in which aesthetics is more than a human ability, playing out between objects as well. Harman answers, "Sure," and he clarifies how aesthetics for him "is about the separation of an object from its qualities" and that it "might be a broader term than art. Aesthetics is the term I use for any separation between an object and its qualities."[134] In this sense, there is correspondence to be found between Bateson's notion of the aesthetic and Harman's. This way, it does make sense to extend—or at least consider to extend—"the artful" to all things. That is, the artful, not as strictly the artistic practice of producing artworks, but as the general outlook on the world that is attuned to forms and relations, in other words, to contexts and their interaction.

To be clear, such an outlook is something that has to be practiced. Learning to become artful is—once more—first and foremost about changing our epistemological glasses, our epistemological lens: learning to see differently. In the first instance, we need to assimilate, incorporate really, this way of looking. This is hard. We don't change our worldview overnight; it has been imprinted on us firmly through culture, morality, education. It will be partly up to philosophers of technology to bring about and ensure that the newly emerging ecological paradigm can become, in the long run, at least a minimal component of the collective consciousness. A few cues can be given for this work. Reorienting and redefining our perspective on technology around the notion of purpose entails thinking about *goals*, on so many levels. We can still learn from domains such as media literacy and technology literacy, but what we need most prominently and most urgently is a *goal literacy*.[135] We need to start thinking differently about goals or, more precisely, we need to *start thinking about goals as such*—since at this moment goals, certainly on a societal level, are scantly thought about, as they are simply not made explicit. They remain lingering, working their effect, subliminally, just beneath the surface of ideological structures. We need to start making goals explicit, talking about them, questioning them, debating them. Where should we have these discussions? On as many levels, in as many contexts as possible: on a societal level, individually, in organizations, companies, governments. The question *"What is it for?"* should at any time, in all these contexts, be at the front of our attention—of course not (just) in the conscious-purposive sense, but seen from the viewpoint of the framework that was constructed here. This means: assessing technology as purpose; and a central way into this is asking about things' purpose, *seeing purposive structures*. Remember my breakfast illustration. We can wonder about the concrete-abstract connections typifying each given situation, or the digital-analog relations at play, or the quantity-quality ratios at stake. Essentially, all these are grounded in investigations of purposive structures—always. My breakfast is a purposive structure; it serves all sorts of purposes, taken up as it is in a network of relations—serving to nourish me, give me energy enough to work, but also allowing me to slowly wake up, get ready for the day; and from there on, it spins out into ever wider circles of purposive orientation. This goes for all objects in the Harmanist sense, all contexts in the Batesonian sense; we can ask about every object's and context's purpose—once again not exclusively in the conscious-purposive sense, and not in classic Aristotelian or religious ways, but in the systemic-wise sense that Bateson outlines.[136] This requires, in the end, an aesthetic, artful sensitivity. We have to become sensitive to the beauty of detours. And to the extent that this sensitivity is aesthetic, it is ethical. How *beautiful* do we find, for instance, the way in which chickens are treated in the meat industry, packed together with thousands in battery cages in gigantic, noisy halls? How aesthetically pleasing do we find catching tuna with drift nets of a couple of kilometers long? If all standard moral reasoning and ethical debate fails, or won't suffice, an aesthetic-artful perspective—in the sense proposed here—might open our eyes to matters of ecological balance and proportion, and offer in itself a detour into fresh assessments (the size of the net is gigantically out of proportion to the fish that needs to be caught, and especially to the bycatch—an aesthetic-moral evaluation). *What is it for?* A hard and simple question at the same time; but the sooner we ask it, the less regret we'll have afterward, because our goals and purpose will have been, at least to *some* extent, appropriated as *ours*.

Concluding Reflections

It might be risky to build up a whole argument around a term such as "purpose." Certainly in the North American context, the word rings with undertones of religion, new age spirituality, and folk psychology. "Find your purpose in life." "Discover your true purpose." At the very least, purpose seems to be merely a private matter of existential meaning-making. But this is exactly my point: we need to get the notion out of those realms alone. Our understanding of technology, of matters of technology, gains by looking at them not only through the lens of purpose, but *as* purpose. On the other hand, our notion of purpose can *also* gain by understanding it *through* looking at real-world systemic structures that involve technology—not merely in the Silicon Valley sense of "technology x serves bigger purpose y and greater good z," but accounting for shared processes of (political) responsibility-taking.

This requires new terms, new words. Persistently referring to technology as if there is such a well-circumscribed "something" harbors danger. "*Something* thinking" diverts us from important issues: what these so-called technologies actually effectuate, what sort of culture they help to install and consolidate, what ways of thinking they open up and close off. To be sure, not in the determinist sense, but in the systemic sense where so many objects, so many contexts are entwined: "we" are all in this together. Neither our common-sense instrumentalism (or Silicon Valley enthusiasm) nor Hollywood dystopianism will do to help us make sense of this. But the same goes for *purpose*: we also need to pursue purpose along the middle road, in between the extremes of narrow instrumental purposiveness and all-encompassing, grand, "given" purpose. Purposes come about in interrelated purposive structures, which doesn't mean they can't be "big." We sometimes need *big* purposes—saving our earthly ecosystem, for one. Walking the purposive middle road involves us (humans) taking responsibility for small as well as bigger purposes; but also incorporating and including "technologies" in this task since, as we saw, they can be and should be made instrumental in achieving systemic wisdom.

As we learn to think differently, see technology not as an *all* or a *this*, but as a *non-all*, a *not just this*, we become more confident in the face of the vanishing thing that technology is. Bizarrely enough, technology *is* this *non-all*, this *not just this*. Rather than a *this* or a *that*, it is a *here and there*. We look out from-inside-to-outside. The paradox in philosophy *of technology* evaporates. The core problems plaguing it (concerning instrumentalism, the empirical turn, the ontological-ontic friction around the human-technology continuity) are cast in a different light. The central dichotomy, leveled down now to countless contexts, does not haunt us anymore with its internal tensions. And instead of exclusively reasoning in terms of linear A-to-B trajectories, we are now able to appreciate the beauty in detours. In the Batesonian sense: we become attentive to the beauty that *is* the ecological interrelation of contexts. A detour is still a *tour*, it has its own purpose: one wouldn't be able to get lost if one weren't moving in a direction. Eventually, though it may take a long time and lots of energy, one can expect to get somewhere. We have to start looking on our detours as full-fledged parts of purpose, not as "waste of time." Like branches stretching out from a trunk to catch the light of the sun, our detours have their own efficiency. But we may not see it that way; we may think that branching out is for us—unlike for trees—a "wrong" thing to do, as it doesn't seem "A-to-B" enough. While, in fact, the detour—what we believe the detour is—is beautiful *because* it is a detour.

We must teach ourselves a new self-image. We are at all times in Maturana and Varela's submarine (see chapter 4). We can develop our conscious purposes, we *must* develop them—we have no other choice; we will never escape the A-to-B principle in this regard—but must remember they come about in, are part of, an external view. At the same time, we need to remind ourselves, we are always inside somewhere, embedded in a network of broader relations, characterized by multipurposiveness. External and internal are different levels. Combined, we are always *in* a context *from* which we look out onto another—possibly with an eye to conquering, controlling, manipulating. So actually we must complete Maturana and Varela's metaphor. It is not that the submarine steersman has an absolutely "pure" view, in contrast to the person on the shore. Yes, the latter has an external standpoint, while the former, the navigator—well practiced and at home in his environment, in his circle—has the internal perspective of spontaneously, without much effort and conscious-purposive thinking, evading the reefs and surfacing (thus, self-producing his system). But even for the navigator, there is *also* an outside that falls out of his circle and is a matter of getting from A to B, "out there." He might surface naturally and easily, but what are his plans on land? Shop for groceries, have a beer, kill someone, fight a war? No one can escape the circle structure, the inside-outside dynamic. Narrow and wider purposes, and levels, are inescapably intertwined.

In this sense, in the philosophy of technology, we *need* different approaches. Verbeek accuses critical perspectives of only talking, of not doing anything (see chapter 6). Talking seems to be indeed an external affair—just standing outside of something and, seemingly noncommittally, commenting upon the thing—while doing is taking the inside view, really making a difference, or so it would appear. But all of the above should attest to the idea that we need *both*; exactly because inside and outside are inseparably linked. They are each other's fate, one might say in a more dramatic way. We require talking and doing—but in the correct proportion. Talking (discourse) and doing (action, matter) are not realms between which we have to choose. Let's talk the talk *and* walk the walk.

And then from there, we go on and imagine ourselves, *our selves*, to be taken up in countless interlocking circles, contexts, objects. Bateson says: "We might say that in creative art man must experience himself—his total self—as a cybernetic model."[137] Despite decades of postmodernism, posthumanism, cyborg theory, and so on, we still have a hard time really *feeling* this. Certainly in regard to current technological developments, such as nanotech, we will in the near future *have* to catch up at one point or another, on the level of our self-image, with these new realities. We will *have* to learn to see ourselves as *also that*. We are—not least in the political sense—no longer the autonomous, unbound unit of Enlightenment discourse. We are, already, something else, "the other"—the countless others of technologies, data, algorithms. At the same time, once more paradoxically, this brings along a responsibility that is indelibly *ours*. *We* will have to come to grips with what we are. This is our challenge to take on. Now how to deal with a changing self? It's almost a therapeutic issue, and in that regard, as far as Bateson is concerned, we should let go of notions such as "discipline" and "self-control." It's no longer about regaining a purity that was lost; it is rather about letting go and trying to embrace, as sensefully as possible, the unknown. We are dealing with "an incorporation or marriage of ideas about the world with ideas about self."[138] Making the jump to a wider context requires a leap of faith[139]—it's a jump into the unknown. It is, indeed, a *jump*.

Where we will wind up is unsure. But we can be certain to land somewhere.

Notes

Preface

1. Bateson, *Steps to an Ecology of Mind*, 512.
2. Foucault, *The Order of Things*, 387.
3. Haraway, *Staying with the Trouble*; Moore, *Anthropocene or Capitalocene?*
4. McLuhan, *Counterblast*, 55.

Introduction

1. See Piore, "The Surgeon Who Wants to Connect You to the Internet with a Brain Implant."
2. See, for instance, Steiner, *Automate This*.
3. See Ihde, *Technology and the Lifeworld*, 4, 26–27; Latour, *Pandora's Hope*, 176ff.
4. Ihde, *Technology and the Lifeworld*, 142.
5. Feenberg, *Questioning Technology*, 193.
6. Verbeek, *What Things Do*.
7. And indeed the spectacle of an atom bomb detonating must inspire such awe that we cannot think much else than "we bit off more than we can chew."
8. See Van Den Eede, "Where Is the Human?" The distinction between ontological and ontic stems from Heidegger and refers to the difference between being and being*s*. Beings, that is, things that *are*, and that we encounter in the world, definitely take part in being, but they *are not being*. Being as such is something deeper. In that vein, human being may be "technological" at heart, but this does not preclude that in everyday life we encounter beings, that is, specific technologies, about which we have to choose whether we want them in our lives (and sometimes literally in our bodies) or not.
9. See Jensen and Rödje, "Introduction," 18ff. Bateson was an important inspiration for Deleuze and Guattari, arguably especially for the latter. See in particular Guattari, *The Three Ecologies*.

10. Heidegger, "The Question Concerning Technology."

11. See Achterhuis, *American Philosophy of Technology*; Kroes and Meijers, *The Empirical Turn in the Philosophy of Technology*.

12. For postphenomenology, see Ihde, *Technology and the Lifeworld*; Verbeek, *What Things Do*. For critical theory of technology, see Feenberg, *Transforming Technology*. For philosophy of information, see Floridi, *The Philosophy of Information*.

13. See Scharff, "Empirical Technoscience Studies in a Comtean World"; Smith, "The Internet as Idea"; Van Den Eede, Goeminne, and Van den Bossche, "The Art of Living with Technology."

14. Lemmens, "Love and Realism"; Smith, "The Internet as Idea"; Smith, "Rewriting the Constitution"; Smith, *Exceptional Technologies*.

15. Smith, "Rewriting the Constitution."

16. Mazlish, *The Fourth Discontinuity*.

17. Verbeek, "Toward a Theory of Technological Mediation," 192.

18. Coeckelbergh, *New Romantic Cyborgs*.

19. See Puech, *The Ethics of Ordinary Technology*.

20. Marshall McLuhan liked to use the term "somnambulist." As one of his biographers, Philip Marchand, describes, "It was a term he would increasingly use [as from around 1960] to characterize those who singlemindedly pursued goals in near total ignorance of what they—and the media they used—were actually doing" (Marchand, *Marshall McLuhan*, 155).

21. See Borgmann, *Technology and the Character of Contemporary Life*; Higgs, Light, and Strong, *Technology and the Good Life?*; Verbeek, *What Things Do*; Brey, Briggle, and Spence, *The Good Life in a Technological Age*.

22. One edited volume in Dutch refers explicitly to "art of living": Hogenhuis and Koelega, *Technologie als levenskunst*. It zooms in on issues such as responsibility, technology assessment, moral consultation, sustainability, participation, and democratization.

23. Feenberg, *Heidegger and Marcuse*, 99.

24. Research, for instance, has been involved with questioning the "object" or "artifact" status of especially digital technology; see, among others, Wiltse, Stolterman, and Redström, "Wicked Interactions"; Soltanzadeh, "Questioning Two Assumptions in the Metaphysics of Technological Objects"; Berry, *Critical Theory and the Digital*; Kember and Zylinska, *Life after New Media*; Bogost, "You Are Already Living Inside a Computer."

Chapter 1

1. Bateson and Bateson, *Angels Fear*, 10; Bateson, *Mind and Nature*, 48.

2. Bateson, *Steps to an Ecology of Mind*; Bateson, *Mind and Nature*, 4; Bateson and Bateson, *Angels Fear*, 9ff.

3. For reasons of scope I will not elaborately sketch Bateson's biography. Material on this, in the form of (intellectual) biographies and/or overviews of his work, is amply available. See Lipset, *Gregory Bateson*; Bateson, *With a Daughter's Eye*; Harries-Jones, *A Recursive Vision*; Charlton, *Understanding Gregory Bateson*; Harries-Jones, *Upside-Down Gods*; Chaney, *Runaway*.

4. *Steps* is a collection of essays and papers by Bateson up to that point in his career. For reasons of clarity, when referring to this work, I will always reference the book as such, not the individual papers. The same goes for the posthumously published Bateson, *A Sacred Unity*.

5. May, "Gregory Bateson and Humanistic Psychology," 84.

6. Quoted in ibid.

7. Bruce Clarke describes how after the Macy Conference the cybernetics movement split into two camps. The "hard" camp "monopolized its resources, hoarded its grants, and redirected the mathematical and engineering sides of cybernetics toward Artificial Intelligence (AI), robotics, computer science, and command-control-communications technologies." The "soft" camp, by contrast, "often loosely identified with the work of Gregory Bateson, gradually gathered up the cognitive and philosophical insights of cybernetics toward matters of managerial and social systems, psychotherapy, and epistemology" (Clarke, "Heinz von Foerster's Demons," 34–35).

8. See also Charlton, *Understanding Gregory Bateson*, 33.

9. Bateson, *Mind and Nature*, 90. Original emphases. Henceforth all emphases in quotes are original, unless otherwise noted.

10. See among others Bateson, *Steps to an Ecology of Mind*, 272.

11. See Mansell, *Imagining the Internet*, 77; Thompson, *Mind in Life*, 57.

12. Bateson, *Mind and Nature*, 90.

13. Bateson and Bateson, *Angels Fear*, 59.

14. Ibid.

15. Ibid.

16. And of course, "social" and "natural" can be seen to correspond to mind and matter, respectively.

17. See Ruesch and Bateson, *Communication*.

18. See Bateson, *Mind and Nature*, 29ff.

19. Bateson, *Steps to an Ecology of Mind*, 450.

20. Hayles, *How We Became Posthuman*.

21. For an overview of the history of cybernetics, see Heylighen and Joslyn, "Cybernetics and Second-Order Cybernetics." Andrew Pickering has written a history of specifically British cybernetics, in which he discusses Bateson—rather briefly, and zooming in mostly on his schizophrenia theory—in a chapter together with R. D. Laing: Pickering, *The Cybernetic Brain*. An early and famous overview of cybernetics is Ashby, *An Introduction to Cybernetics*.

22. Capra, *The Tao of Physics*.

23. Capra, *The Web of Life*.

24. See Strate, *On the Binding Biases of Time and Other Essays on General Semantics and Media Ecology*.

25. Korzybski, "General Semantics, Psychiatry, Psychotherapy and Prevention," 299.

26. Anton, *Communication Uncovered*. Concomitant sources here are Anthony Wilden's *System and Structure* and his later *The Rules Are No Game*, both of which dig deep into Bateson's theory of communication.

27. Bowers et al., *Perspectives on the Ideas of Gregory Bateson, Ecological Intelligence, and Educational Reforms*.

28. Henk Oosterling, *Woorden als daden*.

29. van Boeckel, *At the Heart of Art and Earth*.

30. Flemons, *Completing Distinctions*.

31. Mansell, *Imagining the Internet*, 4.

32. Halpern, *Beautiful Data*.

33. Angus, *Primal Scenes of Communication*.

34. See ibid., 45ff.

35. In Bowers et al., *Perspectives*, there is some discussion of technology, especially in relation to educational uses, but no analysis of technology from a fundamental philosophical perspective, as I seek to offer here. Also, Gianfranco Savino's self-published (and alas not very well written) *Ontology of Complexity* draws parallels between Bateson's ontology and Heidegger's philosophy, but without going into the matter of technology as such.

36. Bateson, *A Sacred Unity*, 313.

37. In the same vein Heidegger conceives the hermeneutic circle in *Being and Time*.

38. Wittgenstein, *Tractatus Logico-Philosophicus*, 27.

39. Bateson and Bateson, *Angels Fear*, 163.

40. Bateson, *Mind and Nature*, 24.

41. Within the book, the contributions by Gregory and Mary Catherine are clearly indicated. In referring to the work, I am discussing the passages written by Gregory, unless otherwise noted.

42. I want to emphasize "idiosyncratic." Indeed, I am offering an idiosyncratic reading here to the extent that I am less directly interested in typical Batesonian themes such as psychology and communication. I want to deploy fundamental lines of force in Bateson's framework in the service of finding a new way of thinking about technology, against the backdrop of philosophy of technology. Such a project by necessity leaves some ground uncovered; I will not go into certain aspects of the vast Batesonian conceptual universe—although I believe I elucidate at least to a minimal degree what I think is essential to it.

Chapter 2

1. Heidegger, "The Question Concerning Technology."

2. Heidegger, "'Only a God Can Save Us.'"

3. See Ihde, *Technology and the Lifeworld*; Verbeek, *What Things Do*; Feenberg, *Transforming Technology*; Winograd and Flores, *Understanding Computers and Cognition*.

4. Heidegger, *Being and Time*, 95ff.

5. For an earlier elaboration of the central dichotomy and its instantiation in several approaches within contemporary philosophy of technology and in related perspectives—albeit from a slightly different perspective, namely, the distinction between transparency and opacity—see Van Den Eede, "In Between Us."

6. For an introduction to postphenomenology, see Rosenberger and Verbeek, "A Field Guide to Postphenomenology."

7. See Ihde, *Technics and Praxis*; Ihde, *Technology and the Lifeworld*; Ihde, *Heidegger's Technologies*.

8. Ihde, *Technology and the Lifeworld*, 72–123.

9. See among others Rosenberger, "The Phenomenological Case for Stricter Regulation of Cell Phones and Driving"; Wellner, "Multi-Attention and the Horcrux Logic"; Rosenberger, "Multistability and the Agency of Mundane Artifacts"; Irwin, *Digital Media*; Wellner, *A Postphenomenological Inquiry of Cell Phones*; Wiltse, "Unpacking Digital Material Mediation"; Van Den Eede, Irwin, and Wellner, *Postphenomenology and Media*.

10. Ihde, *Heidegger's Technologies*, 106.

11. Ihde, *Technics and Praxis*, 128.

12. We will find later that in fact technology *is a sort of all*, albeit of an extremely multifaceted nature, certainly not a "lump," and the *Zuhandenheit-Vorhandenheit* tension can be preserved.

13. Some might not spontaneously categorize Latour as a philosopher of technology; he himself in the first instance would probably not embrace that denomination. But his work is so influential in and essential to the field that, in that capacity, I believe we can and must treat it as belonging to it.

14. Latour, *We Have Never Been Modern*.

15. See Latour, *Science in Action*; Latour, *Aramis*; Latour, *Reassembling the Social*.

16. Pinch and Bijker, "The Social Construction of Facts and Artifacts."

17. Harman, *Prince of Networks*, 37.

18. Latour, *Science in Action*, 131.

19. Feenberg, *Critical Theory of Technology*; Feenberg, *Transforming Technology*; Feenberg, *Questioning Technology*. More recently, Feenberg has begun to describe his approach as "critical constructivism"; see Feenberg, *Technosystem*.

20. Feenberg, "Subversive Rationalization."

21. See Feenberg, *Transforming Technology*, 20–21.

22. Ibid., 21.

23. Ibid.

24. Feenberg, *Technosystem*, 8; Feenberg, *Questioning Technology*, 98.

25. Feenberg, *Questioning Technology*, 95.

26. Ibid., 95, 89.

27. Feenberg, *Heidegger and Marcuse*, 105.
28. Feenberg, *Between Reason and Experience*, xviii.
29. Feenberg, *Alternative Modernity*, 27.
30. Feenberg, *Heidegger and Marcuse*, 109.
31. Feenberg, *Questioning Technology*, 120–29.
32. Quoted in Feenberg, "Modernity Theory and Technology Studies," 82.
33. See Feenberg, *Transforming Technology*.
34. See Feenberg, *Alternative Modernity*.
35. It is of course a bit more complex than represented here. Vestigial subject-object thinking works differently in instrumentalism than in essentialism. In instrumentalism, it is more easily recognizable; subjects and objects come together in a kind of utilitarian atomism. Objects are put in the service of subjects, in order to fulfill their goals. In essentialism, by contrast, an all-encompassing "object" works its influence on countless frail subjects, so to speak. This might seem confusing, if one considers for instance Heidegger's purported essentialism. Wasn't Heidegger intent on destroying the Cartesian legacy? But this is what Ihde comments upon when he regrets that Heidegger lets the modes of presence-at-hand and readiness-to-hand collapse in the later technology analysis. According to postphenomenology, Heidegger betrays his initial project.
36. Feenberg, *Questioning Technology*, 179.
37. See Feenberg, *Heidegger and Marcuse*.
38. Sharon, *Human Nature in an Age of Biotechnology*; Coeckelbergh, *Human Being @ Risk*.
39. Verbeek, *Moralizing Technology*, 47.
40. McLuhan, *Understanding Media*.
41. See McLuhan and Nevitt, *Take Today*.
42. Harman, "The McLuhans and Metaphysics."
43. McLuhan and McLuhan, *Laws of Media*, 63. In a certain way, but a crucial one, this is also the central message of this book.
44. Harman, *Bruno Latour*, 57.
45. See for instance the work of Iris van der Tuin: van der Tuin and Iliadis, "Interview with Iris van der Tuin."
46. Puech, *Homo sapiens technologicus*.
47. Ibid., 21. All translations from this book are mine.
48. Ibid., 61.
49. Ibid., 321.
50. Puech, by the way, is heavily influenced by Eastern philosophy.
51. Son, "Are We Still Pursuing Efficiency?," 49.
52. Perhaps this is an all-too-functionalist way of looking at love, that would upset the more romantically talented—but nonetheless such things pose a real practical challenge to most couples, even if they would not always explicitly admit this.
53. See Gunkel, *Thinking Otherwise*.

54. Even though the world *is* digital to a large extent. This is one of the big ironies of our era: we must cope in the digital world, but cannot seem to do this decently when we "think" only "digitally." Obviously, different meanings of digital are at stake here—although they are intertwined. I return to this in chapters 3 and 7.
55. Van Den Eede, "In Between Us."
56. Verbeek, *What Things Do*, 114.
57. Ibid., 194.
58. See also the example of the speed bump in Verbeek, *Moralizing Technology*, 11.
59. See also Verbeek, "Expanding Mediation Theory."

Chapter 3

1. Then again, as I already briefly suggested in the previous chapter, we will later find that beneath the clear and apparent relationality of Bateson's framework also some kind of *substance* lingers, though not as we know it. It will turn out that not *all* is relation. Here fruitful parallels can be drawn with Graham Harman's object-oriented ontology, a task I undertake in chapter 6.
2. Swaab, *We Are Our Brains*.
3. Bateson, *Mind and Nature*, 81–82.
4. Bateson and Bateson, *Angels Fear*, 90.
5. Bateson, *Mind and Nature*, 82.
6. Bateson, *Steps to an Ecology of Mind*, 487–88.
7. Bateson, *Mind and Nature*, 29.
8. Bateson, *A Sacred Unity*, 227.
9. Bateson and Bateson, *Angels Fear*, 19.
10. Bateson, *Mind and Nature*, 4.
11. In recent decades increasingly common ground has been found between phenomenology and systems thinking or cybernetics as such, for example in the later work of Francisco Varela; for a good overview, see Thompson, *Mind in Life*.
12. May, "Gregory Bateson and Humanistic Psychology," 80. See also Harries-Jones, "Consciousness, Embodiment, and Critique of Phenomenology in the Thought of Gregory Bateson."
13. Bateson, *Mind and Nature*, 12.
14. Ibid., 123.
15. Ibid., 13.
16. Ibid., 86.
17. Ibid., 87.
18. Bateson and Bateson, *Angels Fear*, 18.
19. Ibid., 20.
20. See ibid., 19–20.

21. Ibid., 24.
22. Ibid., 18.
23. Once again, with the proviso that beyond the relationality some substance can—even must—still be suspected, much in the vein of object-oriented ontology. See chapter 6.
24. Bateson, *Mind and Nature*, 124.
25. Ibid.
26. For a very clear exposition of the difference between the two, see the vlog by Anton, *Corey Anton*.
27. Ibid.
28. Bateson and Bateson, *Angels Fear*, 119.
29. Biological explanations come up regularly in Bateson's framework. Given the lack of expertise I have in this domain, I cannot go deeply into this aspect—e.g., inquiring how Bateson's thinking in this regard relates to more contemporary insights—but see for instance Harries-Jones, *Upside-Down Gods*.
30. Bateson and Bateson, *Angels Fear*, 115.
31. Ibid., 119.
32. Bateson, *Mind and Nature*, 41.
33. Ibid., 49.
34. See also ibid.
35. Bateson and Bateson, *Angels Fear*, 117.
36. Korzybski's main work is Korzybski, *Science and Sanity*. For a brief introduction to Korzybski and general semantics, see Strate, "Alfred Korzybski and General Semantics."
37. Korzybski, *Science and Sanity*, 387.
38. He illustrates this theory by way of a diagram that he calls the "Structural Differential." A Google search brings up examples. Or see ibid., 396.
39. In fact, in Korzybski's view, human beings' abstracting capacities, which allow them to hand over information to each following generation, a process which he calls "time-binding," are what makes them human. See Korzybski, *Manhood of Humanity*.
40. See Korzybski, *Science and Sanity*, 58ff.
41. Ibid., 185.
42. See Strate, "Alfred Korzybski and General Semantics," 33.
43. Bateson, *A Sacred Unity*, 309.
44. Bateson, *Mind and Nature*, 28.
45. Ibid.
46. Ibid., 124.
47. This in itself, of course, shouldn't come as a surprise given that, as we saw, Bateson was a pioneer of ecological thinking.
48. Berman, *The Reenchantment of the World*, 145.
49. Korzybski, *Science and Sanity*, 161.

50. Bateson and Bateson, *Angels Fear*, 50.
51. Ibid., 20.
52. Ibid., 164.
53. Bateson, *Mind and Nature*, 77. My emphasis.
54. "Tautology" for Bateson could be seen here as a way of explaining the absence of difference in flux; it is the "*internal consistency* of ideas and processes. But every now and then, the consistency gets torn; the tautology breaks up like the surface of a pond when a stone is thrown into it" (ibid., 194).
55. Bateson and Bateson, *Angels Fear*, 164.
56. Ibid., 162.
57. Ibid., 166.

Chapter 4

1. Specifically for philosophy of technology, see Lemmens, Blok, and Zwier, "Special Issue on the Anthropocene."
2. Aristotle, *Metaphysics*, Book I, Chapter III.
3. See McLuhan and McLuhan, *Laws of Media*; Anton, "McLuhan, Formal Cause, and the Future of Technological Mediation."
4. See Capra, *The Web of Life*; McGilchrist, *The Master and His Emissary*; Snow, *The Two Cultures*; Gadamer, *Truth and Method*.
5. Berman, *The Reenchantment of the World*.
6. See among others Heylighen and Joslyn, "Cybernetics and Second-Order Cybernetics."
7. Rosenblueth, Wiener, and Bigelow, "Behavior, Purpose and Teleology."
8. The notion of goal-seeking, not necessarily overlapping with purpose, would deserve a treatment in itself, as would the related concept of teleonomy, that contrasts with teleology to the extent that it denotes merely apparent goal-directedness, without drawing conclusions about inherent purposefulness. See, among others, Wilden, *The Rules Are No Game*, 78–79.
9. There exists a condition called purpose tremor in which this negative feedback loop reverses into a positive one, causing wild uncontrolled movements.
10. Bateson, *Mind and Nature*, 99.
11. See Varela, Thompson, and Rosch, *The Embodied Mind*; Thompson, *Mind in Life*.
12. Ruesch and Bateson, *Communication*, 182. The book is co-authored, but the passage from which this quote derives was written by Bateson.
13. See among others Bateson, *Mind and Nature*, 82, 119.
14. Ibid., 200.
15. Bateson, *Steps to an Ecology of Mind*, 439–40.
16. Ruesch and Bateson, *Communication*, 183. Same remark as in note 12, above.

17. Bateson, *Steps to an Ecology of Mind*, 446.
18. Ibid., 492.
19. Bateson, *Mind and Nature*, 140.
20. Toulmin, "Afterword: The Charm of the Scout," 363.
21. Bateson, *Mind and Nature*, 50.
22. Ibid., 140.
23. Bateson, *Steps to an Ecology of Mind*, 438.
24. Bateson, *Mind and Nature*, 92.
25. Ibid., 91.
26. Bateson, *Steps to an Ecology of Mind*, 440.
27. See Bateson and Bateson, *Angels Fear*, 91.
28. Bateson, *With a Daughter's Eye*, 238.
29. Ibid., 240.
30. Bateson, *Steps to an Ecology of Mind*, 440.
31. Ibid., 444.
32. Ibid., 438–39.
33. Ibid., 439.
34. Ibid., 450.
35. Charlton, *Understanding Gregory Bateson*, 60.
36. Bateson, *Steps to an Ecology of Mind*, 495ff.
37. Ibid.
38. Ibid., 450.
39. Ibid., 497.
40. Ibid., 450.
41. Ibid., 441.
42. Ibid., 442.
43. Bateson, *A Sacred Unity*, 228.
44. Bateson, *Steps to an Ecology of Mind*, 440.
45. May, "Gregory Bateson and Humanistic Psychology," 91.
46. Bateson, *Steps to an Ecology of Mind*, 500.
47. Bateson, *Mind and Nature*, 7, 10.
48. Bateson, *Steps to an Ecology of Mind*, 440.
49. Bateson, *A Sacred Unity*, 228.
50. Bateson, *Steps to an Ecology of Mind*, 473.
51. See also Bateson, *With a Daughter's Eye*, 240.
52. McLuhan, *Understanding Media*, 73.
53. Bateson, *Mind and Nature*, 163.
54. Ibid., 167.
55. Bateson, *Steps to an Ecology of Mind*, 503.
56. Mary Catherine Bateson, *Our Own Metaphor*, 79. Mary Catherine makes this reflection in the context of a contribution by Ted Schwartz at the conference of which that book is a report; more on this event in chapter 5.
57. Bateson, *Steps to an Ecology of Mind*, 503.

58. See ibid., 501.
59. Clark and Chalmers, "The Extended Mind"; Clark, *Natural-Born Cyborgs*; Logan, *The Extended Mind*.
60. Bateson, *Mind and Nature*, 177.
61. Bateson, "Daddy, Can a Scientist Be Wise?," 66.
62. As just one dramatic illustration of the boons of conscious purpose: at the end of the nineteenth century the world life expectancy at birth was around thirty years. By the end of the twentieth century, it had doubled (and has since been increasing). This is due to things such as the development and use of vaccinations, as well as other treatments for deadly diseases—accomplishments coming forth from conscious purpose, one might say.
63. See Capra, *The Web of Life*.
64. Maturana and Varela, *Autopoiesis and Cognition*, 79.
65. Maturana and Varela, *The Tree of Knowledge*, 75ff.; Maturana and Varela, *Autopoiesis and Cognition*, xxff. The second reference concerns the introduction to *Autopoiesis and Cognition* written by Maturana alone.
66. Maturana and Varela, *Autopoiesis and Cognition*, xviii.
67. Maturana and Varela, *The Tree of Knowledge*, 136–37.
68. Ibid., 136.
69. See Thompson, *Mind in Life*, 140ff.
70. Weber and Varela, "Life after Kant."
71. Kauffman, *Investigations*; Kauffman, *Reinventing the Sacred*.
72. Deacon, *Incomplete Nature*.
73. See Maturana and Varela, *Autopoiesis and Cognition*, 118; Clarke, "Interview with Heinz von Foerster," 29.
74. Maturana and Varela, *Autopoiesis and Cognition*, 70–72.
75. See Luhmann, *Introduction to Systems Theory*.
76. Notwithstanding their possible convergences, Bateson only scantly refers to Maturana and Varela. In one place he mentions them in relation to the concept of recursiveness (calling them mathematicians): Bateson, "Afterword," 242.
77. Berman, *The Reenchantment of the World*.
78. Ibid., 45.
79. Ibid., 40.
80. Ibid., 46.
81. Ibid., 49.
82. Ibid., 51.
83. Ibid., 30.
84. Ibid., 31.
85. Ibid., 40.

Chapter 5

1. Bateson, *Steps to an Ecology of Mind*, 440.

2. Charlton, *Understanding Gregory Bateson*, 5.

3. There are, of course, problems with strictly overlaying these distinctions. It could suggest that systemic wisdom solely corresponds to flux, or to matter. To rigidly argue this would be quite strange. Here, too, ambivalence is in play. Systemic wisdom is more about the interrelation between the two sides and about being aware of it—although there is at the same time the suggestion that the "deeper" layer of primary process and of flux (the layer we've lost touch with) is closer to relation as such. By coupling Bateson to Harman in the next chapters and developing my *circles* metaphor, I hope to clear up this issue to some extent.

4. Flemons, *Completing Distinctions*, 11.
5. Bateson, *With a Daughter's Eye*, 227.
6. Flemons, *Completing Distinctions*, 90.
7. Bateson, *With a Daughter's Eye*, 238.
8. Brand, *II Cybernetic Frontiers*, 17.
9. Flemons, *Completing Distinctions*, 93.
10. Ibid., 89.
11. Ibid., 90.
12. See ibid., 100.
13. Morozov, *To Save Everything, Click Here*.
14. Bateson, *With a Daughter's Eye*, 240.
15. The attribution, however, is disputed, as there seems to exist no original source. See https://en.wikiquote.org/wiki/Margaret_Mead.
16. Bateson, *Steps to an Ecology of Mind*, 444.
17. Quoted in Berman, *The Reenchantment of the World*, 284.
18. Shirky, *Here Comes Everybody*.
19. Kelly, *What Technology Wants*.
20. Morozov, *To Save Everything, Click Here*, 178.
21. Ibid., 220.
22. Bateson, *Our Own Metaphor*.
23. Bateson, *With a Daughter's Eye*, 243.
24. Ibid., 244.
25. Ibid.
26. Bateson, *Our Own Metaphor*, xv.
27. Rorty, *Philosophy and the Mirror of Nature*, 389ff.
28. Bateson, *Our Own Metaphor*, xv.
29. Harries-Jones, *A Recursive Vision*, 50.
30. Quoted in ibid., 49.
31. Bateson, *Steps to an Ecology of Mind*, 438.
32. Bateson, *With a Daughter's Eye*, 244.
33. Ibid., 248.
34. Bateson, *Our Own Metaphor*, 319.

35. Bateson, *With a Daughter's Eye*, 230. Or: "It was in organizational complexity, the relationship between material parts, that he found the matter for almost religious awe" (ibid., 245).

36. Bateson, *A Sacred Unity*, 309.

37. Bateson, *Mind and Nature*, 116.

38. Bateson in several places distinguishes learning from genetics and/or evolution, comparing the characteristics of the two. I will disregard that analysis here, for reasons of scope and focus, and also because for the purposes of the philosophical study of technology, we are chiefly concerned with learning in the sense described throughout this chapter.

39. See Wiener, *The Human Use of Human Beings*, 61.

40. Bateson, *Steps to an Ecology of Mind*, 284.

41. Ibid.

42. Ibid., 293.

43. Bateson, *Mind and Nature*, 13.

44. Bateson, *Steps to an Ecology of Mind*, 301.

45. Ibid., 304.

46. Berman, *The Reenchantment of the World*, 273.

47. Flemons, *Completing Distinctions*, 91.

48. Bateson, *Steps to an Ecology of Mind*, 304.

49. Bateson, *Mind and Nature*, 4. Bateson also briefly mentions the possibility of Learning IV, which "would be *change in Learning III*, but probably does not occur in any adult living organism on this earth" (Bateson, *Steps to an Ecology of Mind*, 293).

50. Bateson, *Steps to an Ecology of Mind*, 287.

51. Ibid., 303.

52. Anton, "Playing with Bateson."

53. See ibid., 39.

54. Bateson, *Mind and Nature*, 164.

55. Bateson, *Steps to an Ecology of Mind*, 301.

56. Ibid., 443, 452, 443, 444, 452.

57. Ibid., 146.

58. Ibid., 147.

59. Quoted in van Boeckel, *At the Heart of Art and Earth*, 108.

60. Just as contemporary philosophy of technology already sees a "more," namely, the network structure of the central dichotomy's wider side. Beyond that "more," however, as we will see, the Batesonian "more" goes even deeper, by enabling us to *multiply* the dichotomy. Technology—what we think technology is—is not *all*, philosophy of technology demonstrates; there is a world of *non-all* out there (see chapter 2). But that *non-all*, we'll find, is just as much "technological," and in that way—paradoxically—again a kind of *all*. See chapters 6 and 7.

61. Bateson, *Steps to an Ecology of Mind*, 272.
62. Bateson, *Mind and Nature*, 133.
63. Ibid., 134.
64. Ibid.
65. Bateson and Bateson, *Angels Fear*, 192.
66. Bateson, *Steps to an Ecology of Mind*, 470.
67. Ibid., 444.
68. Ibid., 445.
69. Ibid., 471.
70. Strate, *On the Binding Biases of Time*, 60. I will discuss Ellul and efficiency more elaborately in chapter 7.
71. Ibid., 61.
72. Ibid.
73. Quoted in ibid.

Chapter 6

1. Wiltse, "Mediating (Infra)structures."
2. McGinn, "What Is Technology?"
3. Ibid., 13.
4. Ibid.
5. Ibid., 23–24.
6. Strangely enough, we might say—as initially Heidegger's tool analysis meant to show how things are enveloped in a network in which one thing refers to another within a purposive dynamic: the hammer refers to the closet that one is making, which refers to the use one wants to make of it to hang one's clothes in and so on. But perhaps afraid to revert to instrumentalism, philosophy of technology has tended to forget this "purpose" part.
7. This—only apparent—ambiguity is showcased beautifully in the amazingly visionary unpublished "Metalogue: Is There a Conspiracy?" from 1971. Bateson's "metalogues" are imaginary, playful conversations between a father and a daughter, partly inspired by real-life conversations between him and Mary Catherine. Notwithstanding their playful character, the dialogues tend to tackle serious questions, and this one in particular has an impressive scope, asking about ecological harm, crisis, destruction, purpose, and technology. Its topic and tone seem today to be more relevant than ever. The father suggests that humans have started to think like "the machines," which are "single-minded." "All machines are shaped rather like mongooses—with sharp noses, and they follow their noses to achieve their purposes." This is the "conspiracy" that the title mentions: humans thinking like machines conspiring with machines. And the conspiracy is aimed toward destruction and ecological damage. Here is a Bateson at his most critical, and perhaps also most

pessimistic. There is a tone of hopelessness: "Crisis does not slow down the process of destruction—*it speeds it up*." To mitigate the harm done by technology we develop other technology, which just tightens the noose. Prodded by his daughter, the father states outright that this process of inventing something with the purpose of solving a specific crisis is "regularly and systematically wrong." The process could not differ more from biological evolution, which does not set out to design with a purpose, and is not obsessed with efficiency. At the end of the dialogue, nevertheless, a door is opened toward the more lenient attitude with regard to technology that Bateson exhibits elsewhere. "Invention" is seen to be not bad per se. The father praises cheese, for example, for its inventive use of resources (though today we would probably point to the ecological damage of cheese production). The metalogue seems to leave us with the conclusion, though left unspoken, that our technological inventions should stay as close as possible to the "natural" ecology. See Bateson, "Metalogue: Is There a Conspiracy?" Rephrased in the vocabulary developed in this book: we should not stop inventing and developing, but rethink that process from the perspective of wider purposive structures, and how these interact with narrow purpose.

8. Latour, *We Have Never Been Modern*; Clark, *Natural-Born Cyborgs*.
9. Borgmann, *Technology and the Character of Contemporary Life*, 197.
10. Ibid., 220.
11. Ibid.
12. Ibid., 221.
13. Ibid., 220. My emphasis.
14. See Verbeek, *What Things Do*.
15. Stiegler, *What Makes Life Worth Living*, 64.
16. See, among others, Stiegler, *The Re-Enchantment of the World*.
17. Ross, "The Ecology of Spirit."
18. Ibid., 11.
19. Ibid., 13.
20. A process that Korzybski calls "time-binding." See chapter 3, note 39.
21. Quoted in Ross, "The Ecology of Spirit," 14. See Bateson, *Steps to an Ecology of Mind*, 492.
22. Ross, "The Ecology of Spirit," 14.
23. Ibid.
24. Stiegler, *Uncontrollable Societies of Disaffected Individuals*, 81.
25. Ibid., 1–126 passim.
26. Ross, "The Ecology of Spirit," 23.
27. Stiegler, *Technics and Time, 1*, 13.
28. Ibid.
29. Ibid.
30. Lemmens, "Love and Realism"; Lemmens, "Thinking through Media."
31. Bateson, *Mind and Nature*, 124.
32. Ibid., 130.

33. See Ihde, *Technology and the Lifeworld*, 42ff.; Ihde, *Husserl's Missing Technologies*, 35ff.; Ihde, *Instrumental Realism*.
34. Ihde, *Postphenomenology and Technoscience*, 55.
35. Ihde, *Bodies in Technology*, 69.
36. See Thompson, *Mind in Life*.
37. Hickman, *John Dewey's Pragmatic Technology*; Hickman, *Philosophical Tools for Technological Culture*.
38. Hickman, *Philosophical Tools for Technological Culture*, 12.
39. Hickman, *John Dewey's Pragmatic Technology*, xii.
40. The notion is ubiquitous in Ihde's work, but for a good exposition and vivid application see Rosenberger, "Multistability and the Agency of Mundane Artifacts."
41. Bateson, *Steps to an Ecology of Mind*, 39–40.
42. Ibid., 40.
43. I leave here undiscussed Latour's more recent *An Inquiry into Modes of Existence*, mostly for the admittedly pragmatic reason that the scope of that book, which is meant as a synthesis of Latour's work up to that point, is so expansive that a substantial treatment of it would almost require a book in itself, or at the least a couple of chapters. A few similarities with my approach could be spotted—if only on a superficial level, for example, Latour also deploys the map-territory scheme—although a hands-on comparison would still be challenging given the different vocabularies. Also, in his even more recent *Facing Gaia*, Latour deploys the imagery, as I do, of multiplying and multiplicity, though coming from another angle (the multiplicity of agency). In any case, despite apparent kinship, Latour does not account for purpose as my approach centrally seeks to do. See Latour, *An Inquiry into Modes of Existence*; Latour, *Facing Gaia*. See also Harman, *Bruno Latour*, 89ff.
44. Marcuse, *One-Dimensional Man*.
45. Ibid., 87.
46. Ibid., 12.
47. Ibid., 97.
48. Ibid.
49. Ibid., 107.
50. Ibid., 79.
51. Ibid., 90.
52. Ibid., 97. My emphasis.
53. Ibid., 107.
54. Ibid., 199.
55. Ibid., 134.
56. Ibid., 180.
57. Verbeek, *What Things Do*; Verbeek, "Expanding Mediation Theory." For a critical assessment of mediation, see Dorrestijn, "The Care of Our Hybrid Selves."
58. See Feenberg, "Modernity Theory and Technology Studies." ANT as such is often criticized for its purported relativism.

59. Verbeek, "Resistance Is Futile." See also Ihde and Selinger, *Chasing Technoscience*, 8.
60. Feenberg, "Peter-Paul Verbeek: Review of 'What Things Do.'"
61. See Van Den Eede, "The Mailman Problem."
62. Rao et al., "Technological Mediation and Power."
63. See Ihde, *Husserl's Missing Technologies*.
64. Smith, "Rewriting the Constitution."
65. For attempts at extending Feenberg's framework, see Van Den Eede, "Extending Feenberg."
66. Verbeek, "The Politics of Technological Mediation." The discussions took place in the context of the 2017 conference of the Society for Philosophy and Technology (SPT) and the 2017 conference of the Society for Social Studies of Science (4S).
67. Verbeek, "Resistance Is Futile."
68. Verbeek, "Accompanying Technology."
69. Feenberg, "What I Said and What I Should Have Said."
70. Feenberg, *Technosystem*, 62.
71. Verbeek in the discussion following his "The Politics of Technological Mediation" talk at SPT 2017 (see above), June 16, 2017.
72. See again Flemons, *Completing Distinctions*.
73. Van Den Eede, "In Between Us."
74. Harman is usually not treated as a philosopher of technology, although some, like Golfo Maggini, refer to him in that way, and he contributed, as also Ihde remarks, a chapter to the important volume *New Waves in Philosophy of Technology*. Others, such as Matt Hayler (to whose work I will shortly turn) and myself, seek to make more clear and elaborate his possible contribution to philosophy of technology. See Maggini, "On the Status of Technology in Heidegger's Being and Time"; Ihde, *Heidegger's Technologies*, 117; Olsen, Selinger, and Riis, *New Waves in Philosophy of Technology*; Hayler, *Challenging the Phenomena of Technology*.
75. Harman, *Prince of Networks*, 141.
76. Harman, *Tool-Being*.
77. Harman, *Heidegger Explained*, 63.
78. Harman, *Immaterialism*.
79. Latour's thought is connected to the new materialist wave, despite his grave reservations about "materialism," that is, materialism as "sound, table-thumping materialism," which understands matter as mathemizable qualities, but is in that sense actually rather idealist (Latour, "Can We Get Our Materialism Back, Please?," 138). See also Harman, *Immaterialism*, 19.
80. Harman, *Immaterialism*, 20. And Harman remarks with Levi Bryant how these new kinds of materialism have not much to do with "anything material" anymore (quoted in *Immaterialism*, 15).
81. Harman, *Prince of Networks*, 187.

82. Harman, "The McLuhans and Metaphysics," 115.
83. Harman, *Prince of Networks*, 142.
84. Harman, *Tool-Being*, 80ff.
85. Harman, *The Quadruple Object*, 20ff.
86. But for an elaboration see Van Den Eede, *Amor Technologiae*, 192–94.
87. Harman, *Bells and Whistles*, 185.
88. See in this regard Bryant, *The Democracy of Objects*, 135–92. Bryant studies autopoietic theory (see chapter 4) within an object-oriented context and especially engages with Luhmann's perspective (briefly referencing in the process Bateson's notion of information as difference that makes a difference), which he sees overlapping with Harman's. "Like Harman's objects, Luhmann's systems are autonomous individuals that are closed and independent of other systems" (*The Democracy of Objects*, 161). Harman himself mentions autopoiesis in a discussion of Latour: Harman, *Bruno Latour*, 105–6.
89. Bateson and Bateson, *Angels Fear*, 18.
90. Bateson, *Mind and Nature*, 178.
91. Harman, *Immaterialism*, 29.
92. Bateson and Bateson, *Angels Fear*, 18.
93. See also this quote by Bateson: "It is necessary to be quite clear about the universal truth that whatever 'things' may be in their pleromatic and thingish world, they can only enter the world of communication and meaning by their names, their qualities and their attributes (that is, by reports of their internal and external relations and interactions)" (Bateson, *Mind and Nature*, 57). We can hear echoes here of Harman's depiction of his fourfold object, constituted by the two crossing dichotomies. The real object stays concealed; we can only relate to objects by way of, indeed, relation.
94. Bateson and Bateson, *Angels Fear*, 20.
95. Harman, *Prince of Networks*, 213. The emphases in this quote testify to Harman's insistence on the connection between psyche and relation, to the extent that objects are not always in a relation, and thus not always "have" psyche. The sentence following the quote goes as follows: "And what I deny is that all entities are always in some sort of relation" (ibid.). Still, the comparison with Bateson remains legitimate, as for Bateson mind is essentially a relational process.
96. Bateson and Bateson, *Angels Fear*, 50.
97. Hayler, *Challenging the Phenomena of Technology*.
98. Also, Hayler's Harman adaptation is further problematic: Hayler argues that an asymptotic movement toward the real object is possible, while Harman denies we can have direct knowledge of it; there is only the possibility of indirect knowledge. But accepting Hayler's premise creates a logical inconsistency within the OOO framework. In keeping with the framework, one would rather have to say that whenever some knowledge is gained, whenever expertise is increased—thus, on a fundamental level, whenever a new relation with something comes about—a

new object emerges. I elaborate this issue in more detail elsewhere (Van Den Eede, "Imagining Things").

99. Harman, *Tool-Being*, 45.
100. Ibid., 67.
101. Harman, *The Quadruple Object*, 42.
102. See Nora Bateson, *Small Arcs of Larger Circles*, in which she deploys a "circles" image to make sense of (Batesonian) contexts. I developed my own imagery of circles independently from this marvelous collection of essays and reflections, but of course clearly the convergence is there, if only because of the shared sources.
103. Harman, *The Quadruple Object*, 119–20.
104. Bateson and Bateson, *Angels Fear*, 164.
105. Ibid., 162. Timothy Morton in his *Hyperobjects*, based for a good part in OOO, remarks helpfully how "our discourse and maps and plans regarding things are not those things. There is an irreducible gap" (Morton, *Hyperobjects*, 133).
106. A sentiment not wholly dissimilar to this is conveyed by Feenberg in *Heidegger and Marcuse* when he describes philosophy of technology, after having traced its origins back to the Greeks, as "the foundation of all Western philosophy" (Feenberg, *Heidegger and Marcuse*, 8).
107. Again, with the proviso made above; see this chapter's note 98.
108. Bateson and Bateson, *Angels Fear*, 187–88.

Chapter 7

1. Bateson, *Mind and Nature*, 128.
2. See Lewis, *Flash Boys*.
3. See Salmon and Stokes, "Algorithms Take Control of Wall Street."
4. Coeckelbergh, *New Romantic Cyborgs*, 158.
5. Floridi, *The Philosophy of Information*, 52.
6. Floridi and Sanders, "On the Morality of Artificial Agents," 354.
7. Floridi, *The Ethics of Information*, 32.
8. Ibid.
9. Ibid.
10. Ibid., 31.
11. Ibid., 32.
12. Ibid., 52.
13. Floridi and Sanders, "On the Morality of Artificial Agents"; Floridi, *The Ethics of Information*.
14. Floridi, *The Ethics of Information*, 33.
15. Ibid.
16. Often attributed to Joseph Heller, specifically his novel *Catch-22*, but the line doesn't seem to appear in the book.

17. Rushkoff, *Program or Be Programmed*, 72.
18. Ibid., 75.
19. Ibid., 80.
20. Ibid., 67.
21. Serres, *Thumbelina*.
22. Ibid., 20.
23. Ibid., 42. The part "they know how to stop at originality" is my own translation. The published English translation reads here "they know how to prevent originality"; but this in my view does not do justice to the original "qu'elles savent s'arrêter à l'originalité" (Serres, *Petite poucette*, 46). My translation is perhaps less eloquent and too literal (it is also the way Google translates the phrase), but truer to the meaning of the French text as far as I'm concerned.
24. Berman, *The Reenchantment of the World*, 141.
25. Ibid., 51.
26. Ibid., 141.
27. See Webster, *Theories of the Information Society*.
28. Ibid., 34.
29. Rushkoff, *Program or Be Programmed*, 84.
30. Wark, *A Hacker Manifesto*, 002.
31. Welzer, *Selbst denken*. All translations from this book are mine.
32. Ibid., 244–45.
33. Pickering, *The Mangle of Practice*.
34. Verbeek, "Toward a Theory of Technological Mediation," 192.
35. Ibid.
36. Coeckelbergh, *Using Words and Things*, 17.
37. Ibid., 13.
38. Ibid., 258.
39. Coeckelbergh, *New Romantic Cyborgs*, 8.
40. Ibid., 9.
41. Smith, *Exceptional Technologies*, 27, 113. Smith however does not directly engage with the "material turn," but names other turns that have been proposed: "engineering turn," "ethical turn," "societal turn," "policy turn," and so on (ibid., 32).
42. Smith, in fact, wants to combine the empirical with the transcendental; both complement each other. His approach in *Exceptional Technologies* exhibits many affinities with mine here. Instead of "turning" as main metaphor, he proposes to speak in terms of "topology," "topography," and "mapping"—as I will later speak of "topologizing." Spatial metaphors, Smith observes astutely, enable thinking "at varying levels of complexity: by expanding and contracting the scope of the space surveyed by zooming in and out" (ibid., 33).
43. See, for instance, Puech's critique of the obsession with discourse and corresponding remedy of sticking mainly with action, but also even to a certain

extent Feenberg's notion of "technological unconscious," which does position discourse and matter as closely intertwined (there are certain interactions between "the conscious" and "the unconscious"), but still suggests that it's *either-or*: values are either discursively available for discussion or embedded in technological design. I fully agree with this evaluation, but I want to put forward and emphasize the idea that the being "there" of both at the same time is a constant; we are always enfolded in discursive-material structures. That means, if we "discursivize" one thing—put it "outside," place it before us—we'll still only be able to do that *from* a certain "material" "inside"; we can never attain a purely discursive or a purely material position. Both spheres and the *gap* in between, in the Batesonian and Harmanist sense, remain.

44. Barad, *Meeting the Universe Halfway*.
45. Ibid., 44–45.
46. Ibid., 33.
47. Ibid., 185.
48. Hayles, *How We Became Posthuman*, 114.
49. Ibid., 115.
50. Ibid., 205.
51. Mansell, *Imagining the Internet*, 178.
52. Ibid.
53. Ibid., 179.
54. Ibid., 76.
55. Ibid., 90. My emphasis.
56. Bateson also seems to jump on this bandwagon when he largely equates purpose (understood as conscious purpose), technology, and efficiency (Bateson, "Metalogue: Is There a Conspiracy?"). Arguing with and against him, however, we could say, "invention" of the kind more akin to biological evolution—the kind that he seems to favor—has its own "efficiency," as will be elaborated further in what follows.
57. Ellul, *The Technological Society*.
58. For an account of instrumental reason, see Angus, *Technique and Enlightenment*.
59. Horkheimer, *Eclipse of Reason*, 102.
60. Feenberg, *Heidegger and Marcuse*, 110.
61. Feenberg and Bakardjieva, "Consumers or Citizens?," 15.
62. Feenberg, *Heidegger and Marcuse*, 103.
63. Ibid.
64. Son, "Are We Still Pursuing Efficiency?," 52.
65. Ibid., 51.
66. See Son, "Are We Still Pursuing Efficiency?"; Alexander, *The Mantra of Efficiency*.
67. Drucker, *Management*, 45.

68. Ritzer, *The McDonaldization of Society*.
69. Boggs, *Phantom Democracy*, 183.
70. Welzer, *Selbst denken*, 110.
71. Stein, *The Cult of Efficiency*, 3, 6.
72. An earlier use of the term "cult" in a discussion on efficiency can be found with Callahan, *Education and the Cult of Efficiency*.
73. Stein, *The Cult of Efficiency*, 7.
74. Alexander, *The Mantra of Efficiency*.
75. Stein, *The Cult of Efficiency*, 16, 17.
76. Morozov, *To Save Everything, Click Here*, 87.
77. Stein, *The Cult of Efficiency*, 19.
78. Bell, *The Coming of Post-Industrial Society*.
79. See Webster, *Theories of the Information Society*, 38ff.
80. Maynard, "Hyping the Efficiencies of Fast(er) Food," 39.
81. Stein, *The Cult of Efficiency*, 26.
82. Alexander, *The Mantra of Efficiency*, 23.
83. Son, "Are We Still Pursuing Efficiency?"; Son, "Reading Jacques Ellul's *The Technological Bluff* in Context."
84. Son, "Are We Still Pursuing Efficiency?," 53.
85. Son, "Reading Jacques Ellul's *The Technological Bluff* in Context," 525.
86. Ibid.
87. Son, "Are We Still Pursuing Efficiency?," 54.
88. Ibid., 57.
89. Stein, *The Cult of Efficiency*, 70.
90. Ibid., 192.
91. Ibid., 219.
92. Son is referring here to Langdon Winner's critical assessment of "autonomous technology." See Winner, *Autonomous Technology*.
93. Son, "Are We Still Pursuing Efficiency?," 58.
94. Ibid., 52, 59–60.
95. Ibid., 60.
96. Ibid.
97. Johnson, "Profiling the Likelihood of Success of Electronic Medical Records," 135.
98. An interesting approach in this light, with affinities to mine—and to Bateson—comes from George Teschner and Alessandro Tomasi, who want to develop an alternative to the notion of efficiency that is elaborated in critiques of Western technology (Ellul, Heidegger, Winner), based on Taoism. In Taoism, action must have its own "efficiency," but in a totally different way. "Truly efficient thinking is an attention to not only the means but to ends as well, that is, it combines efficiency with effectiveness and is able to see usefulness and value in what at first appeared useless and valueless. This ability is due to the flexible, dynamic nature of the Tao,

which effortlessly responds to the specific characteristics of a situation" (Teschner and Tomasi, "Technological Paradigm in Ancient Taoism," 200).

99. The reader who is well acquainted with Bateson's work might ask by now: where are the treatments of Batesonian notions such as "calibration" and "feedback," the sacred (connected in an essential way to the notion of the unconscious), "rigor" and "imagination," play (related to art), and so on? For reasons of scope and feasibility I necessarily have to leave a couple of interesting Batesonian ideas undiscussed—it is almost impossible to do otherwise; his thinking is a treasure trove of insights—but those could, in light of the framework presented here, deserve an elaboration elsewhere.

100. See Shew, *Animal Constructions and Technological Knowledge*.

101. I don't claim this argument as solely my own. In recent times multiple authors have put forward similar proposals, that is, to develop a sensitivity, certainly beyond a user perspective, for the "object perspective." See, for instance, Redström and Wiltse, *Changing Things*.

102. See "How to Be a Systems Thinker: A Conversation with Mary Catherine Bateson," where Mary Catherine Bateson distinguishes between "the computer science side" and "the systems theory side." The interview as such is an excellent primer into the "Batesonian perspective." Also, Mary Catherine offers some interesting reflections on AI, more specifically on the difference between human wisdom and computer intelligence: "One of the most essential elements of human wisdom at its best is humility, knowing that you don't know everything. There's a sense in which we haven't learned how to build humility into our interactions with our devices. The computer doesn't know what it doesn't know, and it's willing to make projections when it hasn't been provided with everything that would be relevant to those projections."

103. Bateson and Bateson, *Angels Fear*, 59.

104. Bateson, "Paradigmatic Conservatism," 353.

105. Ibid.

106. Ibid., 350.

107. Ibid., 351.

108. Ibid., 350.

109. Bateson and Bateson, *Angels Fear*, 189.

110. Berman, *The Reenchantment of the World*, 55–56.

111. Jurgenson and Ritzer, "Efficiency, Effectiveness, and Web 2.0," 53.

112. There might also be a personal preference involved: "It would seem that some people prefer quantitative explanations, while others prefer explanations by the invoking of pattern" (Bateson and Bateson, *Angels Fear*, 115). Also in spite of, or as a correction to, such educationally ingrained or culturally acquired (or perhaps even genetically based) preferences, the two-sided view must be defended.

113. Stein, *The Cult of Efficiency*, 151.

114. Berman, *The Reenchantment of the World*, 252.

115. Ibid., 251.
116. Ibid., 252.
117. Wilden, *The Rules Are No Game*, 225.
118. See "Immosite zegt hoe rijk en gekleurd je buren zijn."
119. Mansell, *Imagining the Internet*, 80.
120. See the introduction.
121. Berman, *The Reenchantment of the World*, 252.
122. Mansell, *Imagining the Internet*, 80, 81.
123. Bateson, *Steps to an Ecology of Mind*, 470.
124. Ibid., 147.
125. Bateson and Bateson, *Angels Fear*, 69.
126. Bateson, *Steps to an Ecology of Mind*, 471.
127. Ibid.
128. Bateson, *Mind and Nature*, 16.
129. Bateson and Bateson, *Angels Fear*, 192.
130. Bateson, *Mind and Nature*, 17.
131. Harman, *The Quadruple Object*, 104.
132. See Harman, *Bells and Whistles*, 66ff.
133. Interesting lectures by Harman in this context, available on YouTube, are *Graham Harman at Moderna Museet: What Is an Object?* and *Graham Harman on Metaphysics, Art, & Speculative Realism*.
134. Harman, *Bells and Whistles*, 221. At the time of this writing, Harman is slated to publish a monograph in late 2019 with Polity on OOO and art, titled *Art + Objects*, which can be expected to go into some of the matters at hand.
135. Van Den Eede, *Vanzelf*.
136. Much like Heidegger in his analysis of "equipment," "assignment" and "involvement" in *Being and Time*, as mentioned, draws a picture of purposive structures. "Involvement" in hammering points to involvement in wanting to make something. This again points to another purpose: perhaps building a shelter, for protection against the weather. And this, then, is for the purpose of Dasein's safety, thus in the end for Dasein as such (see Heidegger, *Being and Time*, 115–17). Indeed, we are back at the original meaning of the tool analysis. Of course, the problematic hint of anthropocentrism lurking here is what Harman argues against.
137. Bateson, *Steps to an Ecology of Mind*, 444.
138. Bateson, *Mind and Nature*, 131.
139. See Harries-Jones, *A Recursive Vision*, 41.

References

Achterhuis, Hans. *American Philosophy of Technology: The Empirical Turn*. Translated by Robert P. Crease. Bloomington: Indiana University Press, 2001.
Alexander, Jennifer Karns. *The Mantra of Efficiency: From Waterwheel to Social Control*. Baltimore: Johns Hopkins University Press, 2008.
Angus, Ian. *Primal Scenes of Communication: Communication, Consumerism, and Social Movements*. Albany: State University of New York Press, 2000.
———. *Technique and Enlightenment: Limits of Instrumental Reason*. Washington, DC: Centre for Advanced Research in Phenomenology & University Press of America, 1984.
Anton, Corey. *Communication Uncovered: General Semantics and Media Ecology*. Fort Worth, TX: Institute of General Semantics, 2011.
———. *Corey Anton: Analog and Digital Communication*, 2010. https://www.youtube.com/watch?v=0T3MbWOK7k0.
———. "McLuhan, Formal Cause, and The Future of Technological Mediation." *Review of Communication* 12, no. 4 (2012): 276–89.
———. "Playing with Bateson: Denotation, Logical Types, and Analog and Digital Communication." In *Communication Uncovered: General Semantics and Media Ecology*, 27–46. Fort Worth, TX: Institute of General Semantics, 2011.
Aristotle. *Metaphysics*. Translated by John H. M'Mahon. London: Henry G. Bohn, 1857.
Ashby, W. Ross. *An Introduction to Cybernetics*. London: Chapman & Hall, 1957.
Barad, Karen. *Meeting the Universe Halfway: Quantum Physics and the Entanglement of Matter and Meaning*. Durham, NC: Duke University Press, 2007.
Bateson, Gregory. *A Sacred Unity: Further Steps to an Ecology of Mind*. Edited by Rodney E. Donaldson. New York: HarperCollins, 1991.
———. "Afterword." In *About Bateson*, edited by John Brockman, 233–47. New York: E. P. Dutton, 1977.
———. "Metalogue: Is There a Conspiracy?" Unpublished manuscript, May 1971. Cited with kind permission of the Bateson Idea Group.
———. *Mind and Nature: A Necessary Unity*. Cresskill, NJ: Hampton Press, 2002.

———. "Paradigmatic Conservatism." In *Rigor & Imagination: Essays from the Legacy of Gregory Bateson*, edited by C. Wilder-Mott and John H. Weakland, 347–55. New York: Praeger, 1981.

———. *Steps to an Ecology of Mind*. Chicago: University of Chicago Press, 2000.

Bateson, Gregory, and Mary Catherine Bateson. *Angels Fear: Towards an Epistemology of the Sacred*. Cresskill, NJ: Hampton Press, 2005.

Bateson, Mary Catherine. "Daddy, Can a Scientist Be Wise?" In *About Bateson*, edited by John Brockman, 55–74. New York: E. P. Dutton, 1977.

———. *Our Own Metaphor: A Personal Account of a Conference on the Effects of Conscious Purpose on Human Adaptation*. Cresskill, NJ: Hampton Press, 2005.

———. *With a Daughter's Eye: A Memoir of Margaret Mead & Gregory Bateson*. New York: Perennial, 2001.

Bateson, Nora. *Small Arcs of Larger Circles: Framing through Other Patterns*. Axminster: Triarchy Press, 2016.

Bell, Daniel. *The Coming of Post-Industrial Society: A Venture in Social Forecasting*. New York: Basic Books, 1999.

Berman, Morris. *The Reenchantment of the World*. Ithaca, NY: Cornell University Press, 1981.

Berry, David M. *Critical Theory and the Digital*. New York: Bloomsbury, 2014.

Boggs, Carl. *Phantom Democracy: Corporate Interests and Political Power in America*. New York: Palgrave Macmillan, 2011.

Bogost, Ian. "You Are Already Living Inside a Computer." *The Atlantic*, September 14, 2017. https://www.theatlantic.com/technology/archive/2017/09/you-are-already-living-inside-a-computer/539193/.

Borgmann, Albert. *Technology and the Character of Contemporary Life: A Philosophical Inquiry*. Chicago: University of Chicago Press, 1984.

Bowers, Chet A., Rolf Jucker, Jorge Ishizawa, and Grimaldo Rengifo. *Perspectives on the Ideas of Gregory Bateson, Ecological Intelligence, and Educational Reforms*. Eugene, OR: Eco-Justice Press, 2011.

Brand, Stewart. *II Cybernetic Frontiers*. New York: Random House, 1974.

Brey, Philip, Adam Briggle, and Edward Spence, eds. *The Good Life in a Technological Age*. London: Routledge, 2014.

Bryant, Levi R. *The Democracy of Objects*. Ann Arbor, MI: Open Humanities Press, 2011.

Callahan, Raymond E. *Education and the Cult of Efficiency: A Study of the Social Forces That Have Shaped the Administration of the Public Schools*. Chicago: University of Chicago Press, 1962.

Capra, Fritjof. *The Tao of Physics: An Exploration of the Parallels between Modern Physics and Eastern Mysticism*. Boston: Shambhala Publications, 1999.

———. *The Web of Life: A New Scientific Understanding of Living Systems*. New York: Anchor Books, 1996.

References

Chaney, Anthony. *Runaway: Gregory Bateson, the Double Bind, and the Rise of Ecological Consciousness*. Chapel Hill, NC: University of North Carolina Press, 2017.
Charlton, Noel G. *Understanding Gregory Bateson: Mind, Beauty, and the Sacred Earth*. Albany: State University of New York Press, 2008.
Clark, Andy. *Natural-Born Cyborgs: Minds, Technologies, and the Future of Human Intelligence*. Oxford: Oxford University Press, 2003.
Clark, Andy, and David Chalmers. "The Extended Mind." *Analysis* 58, no. 1 (1998): 7–19.
Clarke, Bruce. "Heinz von Foerster's Demons: The Emergence of Second-Order Systems Theory." In *Emergence and Embodiment: New Essays on Second-Order Systems Theory*, edited by Bruce Clarke and Mark B. N. Hansen, 34–61. Durham, NC: Duke University Press, 2009.
———. "Interview with Heinz von Foerster." In *Emergence and Embodiment: New Essays on Second-Order Systems Theory*, edited by Bruce Clarke and Mark B. N. Hansen, 26–33. Durham, NC: Duke University Press, 2009.
Coeckelbergh, Mark. *Human Being @ Risk: Enhancement, Technology, and the Evaluation of Vulnerability Transformations*. Dordrecht: Springer, 2013.
———. *New Romantic Cyborgs: Romanticism, Information Technology, and the End of the Machine*. Cambridge, MA: MIT Press, 2017.
———. *Using Words and Things: Language and Philosophy of Technology*. New York: Routledge, 2017.
Deacon, Terrence W. *Incomplete Nature: How Mind Emerged from Matter*. New York: W.W. Norton & Company, 2013.
Dorrestijn, Steven. "The Care of Our Hybrid Selves: Ethics in Times of Technical Mediation." *Foundations of Science* 22, no. 2 (2017): 311–21.
Drucker, Peter F. *Management: Tasks, Responsibilities, Practices*. New York: HarperBusiness, 1993.
Ellul, Jacques. *The Technological Society*. Translated by John Wilkinson. New York: Vintage Books, 1964.
Feenberg, Andrew. *Alternative Modernity: The Technical Turn in Philosophy and Social Theory*. Berkeley: University of California Press, 1995.
———. *Between Reason and Experience*. Cambridge, MA: MIT Press, 2010.
———. *Critical Theory of Technology*. Oxford: Oxford University Press, 1991.
———. *Heidegger and Marcuse: The Catastrophe and Redemption of History*. New York: Routledge, 2005.
———. "Modernity Theory and Technology Studies: Reflections on Bridging the Gap." In *Modernity and Technology*, edited by Thomas J. Misa, Philip Brey, and Andrew Feenberg, 73–104. Cambridge, MA: MIT Press, 2003.
———. "Peter-Paul Verbeek: Review of 'What Things Do.'" *Human Studies* 32, no. 2 (2009): 225–28.
———. *Questioning Technology*. London: Routledge, 1999.

———. "Subversive Rationalization: Technology, Power, and Democracy." In *Technology and the Politics of Knowledge*, edited by Andrew Feenberg and Alastair Hannay, 3–22. Bloomington: Indiana University Press, 1995.

———. *Technosystem: The Social Life of Reason*. Cambridge, MA: Harvard University Press, 2017.

———. *Transforming Technology: A Critical Theory Revisited*. Oxford: Oxford University Press, 2002.

———. "What I Said and What I Should Have Said: On Critical Theory of Technology." *Techné: Research in Philosophy and Technology* 17, no. 1 (2013): 163–78.

Feenberg, Andrew, and Maria Bakardjieva. "Consumers or Citizens? The Online Community Debate." In *Community in the Digital Age: Philosophy and Practice*, edited by Andrew Feenberg and Darin Barney, 1–28. Lanham, MD: Rowman & Littlefield, 2004.

Flemons, Douglas G. *Completing Distinctions: Interweaving the Ideas of Gregory Bateson and Taoism into a Unique Approach to Therapy*. Boston: Shambhala, 1991.

Floridi, Luciano. *The Ethics of Information*. Oxford: Oxford University Press, 2013.

———. *The Philosophy of Information*. Oxford: Oxford University Press, 2011.

Floridi, Luciano, and J. W. Sanders. "On the Morality of Artificial Agents." *Minds and Machines* 14, no. 3 (2004): 349–79.

Foucault, Michel. *The Order of Things: An Archaeology of the Human Sciences*. New York: Vintage Books, 1994.

Gadamer, Hans-Georg. *Truth and Method*. Translated by Joel Weinsheimer and Donald G. Marshall. London: Bloomsbury, 2004.

Graham Harman at Moderna Museet: What Is an Object?, Moderna Museet, 2015. https://www.youtube.com/watch?v=9eiv-rQw1lc.

Graham Harman on Metaphysics, Art, & Speculative Realism, Philosophy Overdose, 2013. https://www.youtube.com/watch?v=ck-fRgNUOAs.

Guattari, Félix. *The Three Ecologies*. Translated by Ian Pindar and Paul Sutton. London: Bloomsbury, 2014.

Gunkel, David J. *Thinking Otherwise: Philosophy, Communication, Technology*. West Lafayette, IN: Purdue University Press, 2007.

Halpern, Orit. *Beautiful Data: A History of Vision and Reason since 1945*. Durham, NC: Duke University Press, 2014.

Haraway, Donna J. *Staying with the Trouble: Making Kin in the Chthulucene*. Durham, NC: Duke University Press, 2016.

Harman, Graham. *Bells and Whistles: More Speculative Realism*. Winchester: Zero Books, 2013.

———. *Bruno Latour: Reassembling the Political*. London: Pluto Press, 2014.

———. *Heidegger Explained: From Phenomenon to Thing*. Chicago: Open Court, 2007.

———. *Immaterialism: Objects and Social Theory*. Cambridge: Polity, 2016.

———. *Prince of Networks: Bruno Latour and Metaphysics*. Melbourne: re.press, 2009.

———. "The McLuhans and Metaphysics." In *New Waves in Philosophy of Technology*, edited by Jan Kyrre Berg Olsen, Evan Selinger, and Søren Riis, 100–122. Basingstoke: Palgrave Macmillan, 2009.

———. *The Quadruple Object*. Winchester: Zero Books, 2011.

———. *Tool-Being: Heidegger and the Metaphysics of Objects*. Chicago: Open Court, 2002.

Harries-Jones, Peter. *A Recursive Vision: Ecological Understanding and Gregory Bateson*. Toronto: University of Toronto Press, 1995.

———. "Consciousness, Embodiment, and Critique of Phenomenology in the Thought of Gregory Bateson." *The American Journal of Semiotics* 19, no. 1/4 (2003): 69–94.

———. *Upside-Down Gods: Gregory Bateson's World of Difference*. New York: Fordham University Press, 2016.

Hayler, Matt. *Challenging the Phenomena of Technology: Embodiment, Expertise, and Evolved Knowledge*. Basingstoke: Palgrave Macmillan, 2015.

Hayles, N. Katherine. *How We Became Posthuman: Virtual Bodies in Cybernetics, Literature, and Informatics*. Chicago: University of Chicago Press, 1999.

Heidegger, Martin. *Being and Time*. Translated by John Macquarrie and Edward Robinson. Oxford: Blackwell, 1962.

———. "'Only a God Can Save Us': The Spiegel Interview (1966)." In *Heidegger: The Man and the Thinker*, edited by Thomas Sheehan, translated by William J. Richardson, 45–67. Abingdon: Routledge, 2017.

———. "The Question Concerning Technology." In *The Question Concerning Technology and Other Essays*, translated by William Lovitt, 3–35. New York: Harper Perennial, 1977.

Heylighen, Francis, and Cliff Joslyn. "Cybernetics and Second-Order Cybernetics." In *Encyclopedia of Physical Science & Technology*, edited by Robert A. Meyers. New York: Academic Press, 2001. http://pespmc1.vub.ac.be/papers/cybernetics-epst.pdf.

Hickman, Larry A. *John Dewey's Pragmatic Technology*. Bloomington: Indiana University Press, 1992.

———. *Philosophical Tools for Technological Culture: Putting Pragmatism to Work*. Bloomington: Indiana University Press, 2001.

Higgs, Eric, Andrew Light, and David Strong, eds. *Technology and the Good Life?* Chicago: University of Chicago Press, 2000.

Hogenhuis, Christiaan, and Dick Koelega, eds. *Technologie als levenskunst: Visies op instrumenten voor inclusieve technologie-ontwikkeling*. Kampen: Uitgeverij Kok, 1996.

Horkheimer, Max. *Eclipse of Reason*. London: Continuum, 2004.

"How to Be a Systems Thinker: A Conversation with Mary Catherine Bateson." *Edge.org*, April 17, 2018. https://www.edge.org/conversation/mary_catherine_bateson-how-to-be-a-systems-thinker.

Ihde, Don. *Bodies in Technology*. Minneapolis: University of Minnesota Press, 2002.
———. *Heidegger's Technologies: Postphenomenological Perspectives*. New York: Fordham University Press, 2010.
———. *Husserl's Missing Technologies*. New York: Fordham University Press, 2016.
———. *Instrumental Realism: The Interface between Philosophy of Science and Philosophy of Technology*. Bloomington: Indiana University Press, 1991.
———. *Postphenomenology and Technoscience: The Peking University Lectures*. Albany: State University of New York Press, 2009.
———. *Technics and Praxis: A Philosophy of Technology*. Dordrecht: D. Reidel Publishing Co., 1979.
———. *Technology and the Lifeworld: From Garden to Earth*. Bloomington: Indiana University Press, 1990.
Ihde, Don, and Evan Selinger, eds. *Chasing Technoscience: Matrix for Materiality*. Bloomington: Indiana University Press, 2003.
"Immosite zegt hoe rijk en gekleurd je buren zijn." *De Morgen*, June 17, 2015. https://www.demorgen.be/economie/immosite-zegt-hoe-rijk-en-gekleurd-je-buren-zijn-b65b2dfd/.
Irwin, Stacey O'Neal. *Digital Media: Human–Technology Connection*. Lanham, MD: Lexington Books, 2016.
Jensen, Casper Bruun, and Kjetil Rödje. "Introduction." In *Deleuzian Intersections: Science, Technology, Anthropology*, edited by Casper Bruun Jensen and Kjetil Rödje, 1–35. New York: Berghahn Books, 2010.
Johnson, J. David. "Profiling the Likelihood of Success of Electronic Medical Records." In *The Culture of Efficiency: Technology in Everyday Life*, edited by Sharon Kleinman, 124–41. New York: Peter Lang, 2009.
Jurgenson, Nathan, and George Ritzer. "Efficiency, Effectiveness, and Web 2.0." In *The Culture of Efficiency: Technology in Everyday Life*, edited by Sharon Kleinman, 51–67. New York: Peter Lang, 2009.
Kauffman, Stuart A. *Investigations*. Oxford: Oxford University Press, 2000.
———. *Reinventing the Sacred: A New View of Science, Reason, and Religion*. New York: Basic Books, 2008.
Kelly, Kevin. *What Technology Wants*. New York: Viking, 2011.
Kember, Sarah, and Joanna Zylinska. *Life after New Media: Mediation as a Vital Process*. Cambridge, MA: MIT Press, 2012.
Korzybski, Alfred. "General Semantics, Psychiatry, Psychotherapy and Prevention." In *Alfred Korzybski: Collected Writings 1920–1950*, edited by M. Kendig, 297–308. Englewood, NJ: Institute of General Semantics, 1990.
———. *Manhood of Humanity*. New York: Institute of General Semantics, 1950.
———. *Science and Sanity: An Introduction to Non-Aristotelian Systems and General Semantics*. New York: Institute of General Semantics, 1994.
Kroes, Peter, and Anthonie Meijers, eds. *The Empirical Turn in the Philosophy of Technology*. Amsterdam: JAI, 2000.

Latour, Bruno. *Facing Gaia: Eight Lectures on the New Climatic Regime*. Translated by Catherine Porter. Cambridge: Polity, 2017.
———. *An Inquiry into Modes of Existence: An Anthropology of the Moderns*. Translated by Catherine Porter. Cambridge, MA: Harvard University Press, 2013.
———. *Aramis or the Love of Technology*. Translated by Catherine Porter. Cambridge, MA: Harvard University Press, 1996.
———. "Can We Get Our Materialism Back, Please?" *Isis: A Journal of the History of Science Society* 98, no. 1 (2007): 138–142.
———. *Pandora's Hope: Essays on the Reality of Science Studies*. Cambridge, MA: Harvard University Press, 1999.
———. *Reassembling the Social: An Introduction to Actor-Network-Theory*. Oxford: Oxford University Press, 2005.
———. *Science in Action: How to Follow Scientists and Engineers through Society*. Cambridge, MA: Harvard University Press, 1987.
———. *We Have Never Been Modern*. Translated by Catherine Porter. Cambridge, MA: Harvard University Press, 1993.
Lemmens, Pieter. "Love and Realism." *Foundations of Science* 22, no. 2 (2017): 305–10.
———. "Thinking through Media: Stieglerian Remarks on a Possible Postphenomenology of Media." In *Postphenomenology and Media: Essays on Human–Media–World Relations*, edited by Yoni Van Den Eede, Stacey O'Neal Irwin, and Galit Wellner, 185–206. Lanham. MD: Lexington Books, 2017.
Lemmens, Pieter, Vincent Blok, and Jochem Zwier, eds. "Special Issue on the Anthropocene." *Techné: Research in Philosophy and Technology* 21, no. 2/3 (2017).
Lewis, Michael. *Flash Boys: A Wall Street Revolt*. New York: W.W. Norton & Company, 2015.
Lipset, David. *Gregory Bateson: The Legacy of a Scientist*. Boston: Beacon Press, 1982.
Logan, Robert K. *The Extended Mind: The Emergence of Language, the Human Mind, and Culture*. Toronto: University of Toronto Press, 2007.
Luhmann, Niklas. *Introduction to Systems Theory*. Translated by Peter Gilgen. Cambridge: Polity Press, 2013.
Maggini, Golfo. "On the Status of Technology in Heidegger's Being and Time." *Studia Philosophiae Christianae* 50, no. 1 (2014): 79–110.
Mansell, Robin. *Imagining the Internet: Communication, Innovation, and Governance*. Oxford: Oxford University Press, 2012.
Marchand, Philip. *Marshall McLuhan: The Medium and the Messenger*. Cambridge, MA: MIT Press, 1998.
Marcuse, Herbert. *One-Dimensional Man: Studies in the Ideology of Advanced Industrial Society*. Boston: Beacon Press, 1966.
Maturana, Humberto R., and Francisco J. Varela. *Autopoiesis and Cognition: The Realization of the Living*. Dordrecht: D. Reidel Publishing Co., 1980.

———. *The Tree of Knowledge: The Biological Roots of Human Understanding.* Boston: Shambhala, 1998.

May, Rollo. "Gregory Bateson and Humanistic Psychology." In *About Bateson*, edited by John Brockman, 75–99. New York: E. P. Dutton, 1977.

Maynard, Michael L. "Hyping the Efficiencies of Fast(er) Food: The Glocalization of McDonald's Snack Wrap in Japan." In *The Culture of Efficiency: Technology in Everyday Life*, edited by Sharon Kleinman, 39–50. New York: Peter Lang, 2009.

Mazlish, Bruce. *The Fourth Discontinuity: The Co-Evolution of Humans and Machines.* New Haven, CT: Yale University Press, 1993.

McGilchrist, Iain. *The Master and His Emissary: The Divided Brain and the Making of the Western World.* New Haven, CT: Yale University Press, 2010.

McGinn, Robert E. "What Is Technology?" In *Technology as a Human Affair*, edited by Larry A. Hickman, 10–25. New York: McGraw-Hill, 1990.

McLuhan, Marshall. *Counterblast.* Designed by Harley Parker. London: Rapp & Whiting, 1970.

———. *Understanding Media: The Extensions of Man.* Corte Madera: Gingko Press, 2003.

McLuhan, Marshall, and Eric McLuhan. *Laws of Media: The New Science.* Toronto: University of Toronto Press, 1988.

McLuhan, Marshall, and Barrington Nevitt. *Take Today: The Executive as Dropout.* New York: Harcourt Brace Jovanovich, 1972.

Moore, Jason W., ed. *Anthropocene or Capitalocene?: Nature, History, and the Crisis of Capitalism.* Oakland: PM Press, 2016.

Morozov, Evgeny. *To Save Everything, Click Here: Technology, Solutionism, and the Urge to Fix Problems That Don't Exist.* London: Allen Lane, 2013.

Morton, Timothy. *Hyperobjects: Philosophy and Ecology after the End of the World.* Minneapolis: University of Minnesota Press, 2013.

Olsen, Jan Kyrre Berg, Evan Selinger, and Søren Riis, eds. *New Waves in Philosophy of Technology.* Basingstoke: Palgrave Macmillan, 2009.

Oosterling, Henk. *Woorden als daden. Rotterdam Vakmanstad/Skillcity 2007–2009.* Heijningen: Jap Sam Books, 2009.

Pickering, Andrew. *The Cybernetic Brain: Sketches of Another Future.* Chicago: University of Chicago Press, 2010.

———. *The Mangle of Practice: Time, Agency, and Science.* Chicago: University of Chicago Press, 1995.

Pinch, Trevor J., and Wiebe E. Bijker. "The Social Construction of Facts and Artifacts: Or How the Sociology of Science and the Sociology of Technology Might Benefit Each Other." In *The Social Construction of Technological Systems: New Directions in the Sociology and History of Technology*, edited by Wiebe E. Bijker, Thomas P. Hughes, and Trevor Pinch, 11–44. Cambridge, MA: MIT Press, 2012.

Piore, Adam. "The Surgeon Who Wants to Connect You to the Internet with a Brain Implant." *MIT Technology Review*, November 30, 2017. https://www.technology

review.com/s/609232/the-surgeon-who-wants-to-connect-you-to-the-internet-with-a-brain-implant/.
Puech, Michel. *Homo sapiens technologicus: Philosophie de la technologie contemporaine, philosophie de la sagesse contemporaine*. Paris: Le Pommier, 2008.
———. *The Ethics of Ordinary Technology*. New York: Routledge, 2016.
Rao, Mithun Bantwal, Joost Jongerden, Pieter Lemmens, and Guido Ruivenkamp. "Technological Mediation and Power: Postphenomenology, Critical Theory, and Autonomist Marxism." *Philosophy & Technology* 28, no. 3 (2015): 449–74.
Redström, Johan, and Heather Wiltse. *Changing Things: The Future of Objects in a Digital World*. London: Bloomsbury Visual Arts, 2019.
Ritzer, George. *The McDonaldization of Society: 20th Anniversary Edition*. Los Angeles: SAGE, 2013.
Rorty, Richard. *Philosophy and the Mirror of Nature*. Oxford: Blackwell, 1980.
Rosenberger, Robert. "Multistability and the Agency of Mundane Artifacts: From Speed Bumps to Subway Benches." *Human Studies* 37, no. 3 (2014): 369–92.
———. "The Phenomenological Case for Stricter Regulation of Cell Phones and Driving." *Techné: Research in Philosophy and Technology* 18, no. 1–2 (2014): 20–47.
Rosenberger, Robert, and Peter-Paul Verbeek. "A Field Guide to Postphenomenology." In *Postphenomenological Investigations: Essays on Human–Technology Relations*, edited by Robert Rosenberger and Peter-Paul Verbeek, 9–41. Lanham, MD: Lexington Books, 2015.
Rosenblueth, Arturo, Norbert Wiener, and Julian Bigelow. "Behavior, Purpose and Teleology." *Philosophy of Science* 10, no. 1 (1943): 18–24.
Ross, Daniel. "The Ecology of Spirit: Technics and Politics in Bernard Stiegler." Unpublished manuscript, 2008. http://www.academia.edu/12685422/The_Ecology_of_Spirit_Technics_and_Politics_in_Bernard_Stiegler_2008_.
Ruesch, Jurgen, and Gregory Bateson. *Communication: The Social Matrix of Psychiatry*. New Brunswick, NJ: Transaction Publishers, 2008.
Rushkoff, Douglas. *Program or Be Programmed: Ten Commands for a Digital Age*. Berkeley, CA: Soft Skull Press, 2011.
Salmon, Felix, and Jon Stokes. "Algorithms Take Control of Wall Street." *WIRED*, December 27, 2010. http://www.wired.com/2010/12/ff_ai_flashtrading/.
Savino, Gianfranco. *Ontology of Complexity: A Reading of Gregory Bateson*. Amazon Kindle book, 2014.
Scharff, Robert C. "Empirical Technoscience Studies in a Comtean World: Too Much Concreteness?" *Philosophy & Technology* 25, no. 2 (2012): 153–77.
Serres, Michel. *Petite poucette*. Paris: Le Pommier, 2012.
———. *Thumbelina: The Culture and Technology of Millennials*. Translated by Daniel W. Smith. London: Rowman & Littlefield International, 2015.
Sharon, Tamar. *Human Nature in an Age of Biotechnology: The Case for Mediated Posthumanism*. Dordrecht: Springer, 2014.

Shew, Ashley. *Animal Constructions and Technological Knowledge*. Lanham, MD: Lexington Books, 2017.

Shirky, Clay. *Here Comes Everybody: The Power of Organizing without Organizations*. London: Allen Lane, 2008.

Smith, Dominic. *Exceptional Technologies: A Continental Philosophy of Technology*. London: Bloomsbury, 2018.

———. "Rewriting the Constitution: A Critique of 'Postphenomenology.'" *Philosophy & Technology* 28, no. 4 (2015): 533–51.

———. "The Internet as Idea: For a Transcendental Philosophy of Technology." *Techné: Research in Philosophy and Technology* 19, no. 3 (2015): 381–410.

Snow, C. P. *The Two Cultures*. Cambridge: Cambridge University Press, 1998.

Soltanzadeh, Sadjad. "Questioning Two Assumptions in the Metaphysics of Technological Objects." *Philosophy & Technology* 29, no. 2 (2016): 127–35.

Son, Wha-Chul. "Are We Still Pursuing Efficiency? Interpreting Jacques Ellul's Efficiency Principle." In *Jacques Ellul and the Technological Society in the 21st Century*, edited by Helena M. Jerónimo, José Luís Garcia, and Carl Mitcham, 49–62. Dordrecht: Springer, 2013.

———. "Reading Jacques Ellul's *The Technological Bluff* in Context." *Bulletin of Science, Technology & Society* 24, no. 6 (2004): 518–33.

Stein, Janice Gross. *The Cult of Efficiency*. Toronto: House of Anansi Press, 2002.

Steiner, Christopher. *Automate This: How Algorithms Took Over Our Markets, Our Jobs, and the World*. New York: Portfolio, 2012.

Stiegler, Bernard. *Technics and Time, 1: The Fault of Epimetheus*. Translated by Richard Beardsworth and George Collins. Stanford, CA: Stanford University Press, 1998.

———. *The Re-Enchantment of the World: The Value of Spirit Against Industrial Populism*. Translated by Trevor Arthur. London: Bloomsbury, 2014.

———. *Uncontrollable Societies of Disaffected Individuals: Disbelief and Discredit, Volume 2*. Translated by Daniel Ross. Cambridge: Polity, 2013.

———. *What Makes Life Worth Living: On Pharmacology*. Translated by Daniel Ross. Cambridge: Polity, 2013.

Strate, Lance. "Alfred Korzybski and General Semantics." In *On the Binding Biases of Time and Other Essays on General Semantics and Media Ecology*, 13–36. Fort Worth, TX: Institute of General Semantics, 2011.

———. *On the Binding Biases of Time and Other Essays on General Semantics and Media Ecology*. Fort Worth, TX: Institute of General Semantics, 2011.

Swaab, D. F. *We Are Our Brains: A Neurobiography of the Brain, from the Womb to Alzheimer's*. Translated by Jane Hedley-Prôle. New York: Spiegel & Grau, 2014.

Teschner, George, and Alessandro Tomasi. "Technological Paradigm in Ancient Taoism." *Techné: Research in Philosophy and Technology* 13, no. 3 (2009): 190–205.

Thompson, Evan. *Mind in Life: Biology, Phenomenology, and the Sciences of Mind*. Cambridge, MA: Belknap Press of Harvard University Press, 2007.

Toulmin, Stephen. "Afterword: The Charm of the Scout." In *Rigor & Imagination: Essays from the Legacy of Gregory Bateson*, edited by C. Wilder-Mott and John H. Weakland, 357–68. New York: Praeger, 1981.
van Boeckel, Jan. *At the Heart of Art and Earth: An Exploration of Practices in Arts-Based Environmental Education*. PhD dissertation. Helsinki: Aalto University/Aalto ARTS Books, 2013.
Van Den Eede, Yoni. *Amor Technologiae: Marshall McLuhan as Philosopher of Technology—Toward a Philosophy of Human-Media Relationships*. Brussels: VUBPRESS, 2012.
———, ed. "Extending Feenberg: Toward the Instrumentalization of the Critical Theory of Technology—Special Issue on Andrew Feenberg's 'Critical Theory of Technology.'" *Techné: Research in Philosophy and Technology* 17, no. 1 (2013).
———. "Imagining Things: Unfolding the 'of' in Philosophy of Technology, through Object-Oriented Ontology." In *Relating to Things: Design, Technology and the Artificial*, edited by Heather Wiltse. London: Bloomsbury, forthcoming.
———. "In Between Us: On the Transparency and Opacity of Technological Mediation." *Foundations of Science* 16, no. 2–3 (2011): 139–59.
———. "The Mailman Problem: Complementing Critical Theory of Technology by Way of Media Theory." *Techné: Research in Philosophy and Technology* 17, no. 1 (2013): 144–62.
———. *Vanzelf: Tegen het efficiëntiedenken en de doelmatigheidscultuur*. Leuven: Acco, 2015.
———. "Where Is the Human? Beyond the Enhancement Debate." *Science, Technology, & Human Values* 40, no. 1 (2015): 149–62.
Van Den Eede, Yoni, Gert Goeminne, and Marc Van den Bossche. "The Art of Living with Technology: Turning over Philosophy of Technology's Empirical Turn." *Foundations of Science* 22, no. 2 (2017): 235–46.
Van Den Eede, Yoni, Stacey O'Neal Irwin, and Galit Wellner, eds. *Postphenomenology and Media: Essays on Human–Media–World Relations*. Lanham, MD: Lexington Books, 2017.
van der Tuin, Iris, and Andrew Iliadis. "Interview with Iris van der Tuin." *Figure/Ground*, December 29, 2014. http://figureground.org/interview-with-iris-van-der-tuin/.
Varela, Francisco J., Evan Thompson, and Eleanor Rosch. *The Embodied Mind: Cognitive Science and Human Experience*. Cambridge, MA: MIT Press, 1991.
Verbeek, Peter-Paul. "Accompanying Technology: Philosophy of Technology after the Ethical Turn." *Techné: Research in Philosophy and Technology* 14, no. 1 (2010): 49–54.
———. "Expanding Mediation Theory." *Foundations of Science* 17, no. 4 (2012): 391–95.
———. *Moralizing Technology: Understanding and Designing the Morality of Things*. Chicago: University of Chicago Press, 2011.

———. "Resistance Is Futile: Toward a Non-Modern Democratization of Technology." *Techné: Research in Philosophy and Technology* 17, no. 1 (2013): 72–92.

———. "The Politics of Technological Mediation." Conference presentation, SPT 2017: The Grammar of Things—Conference of the Society for Philosophy and Technology, Darmstadt, June 16, 2017.

———. "Toward a Theory of Technological Mediation: A Program for Postphenomenological Research." In *Technoscience and Postphenomenology: The Manhattan Papers*, edited by Jan Kyrre Berg O. Friis and Robert P. Crease, 189–204. Lanham, MD: Lexington Books, 2015.

———. *What Things Do: Philosophical Reflections on Technology, Agency, and Design*. Translated by Robert P. Crease. University Park: Pennsylvania State University Press, 2005.

Wark, McKenzie. *A Hacker Manifesto*. Cambridge, MA: Harvard University Press, 2004.

Weber, Andreas, and Francisco J. Varela. "Life after Kant: Natural Purposes and the Autopoietic Foundations of Biological Individuality." *Phenomenology and the Cognitive Sciences* 1, no. 2 (2002): 97–125.

Webster, Frank. *Theories of the Information Society*. 4th edition. London: Routledge, 2014.

Wellner, Galit. *A Postphenomenological Inquiry of Cell Phones: Genealogies, Meanings, and Becoming*. Lanham, MD: Lexington Books, 2016.

———. "Multi-Attention and the Horcrux Logic: Justifications for Talking on the Cell Phone While Driving." *Techné: Research in Philosophy and Technology* 18, no. 1/2 (2014): 48–73.

Welzer, Harald. *Selbst denken: Eine Anleitung zum Widerstand*. Frankfurt am Main: S. Fischer, 2013.

Wiener, Norbert. *The Human Use of Human Beings: Cybernetics and Society*. Boston: Da Capo Press, 1988.

Wilden, Anthony. *System and Structure: Essays in Communication and Exchange*. London: Tavistock Publications, 1972.

———. *The Rules Are No Game: The Strategy of Communication*. London: Routledge & Kegan Paul, 1987.

Wiltse, Heather. "Mediating (Infra)structures: Technology, Media, Environment." In *Postphenomenology and Media: Essays on Human–Media–World Relations*, edited by Yoni Van Den Eede, Stacey O'Neal Irwin, and Galit Wellner, 3–25. Lanham, MD: Lexington Books, 2017.

———. "Unpacking Digital Material Mediation." *Techné: Research in Philosophy and Technology* 18, no. 3 (2014): 154–82.

Wiltse, Heather, Erik Stolterman, and Johan Redström. "Wicked Interactions: (On the Necessity of) Reframing the 'Computer' in Philosophy and Design." *Techné: Research in Philosophy and Technology* 19, no. 1 (2015): 26–49.

Winner, Langdon. *Autonomous Technology: Technics-out-of-Control as a Theme in Political Thought*. Cambridge, MA: MIT Press, 1977.
Winograd, Terry, and Fernando Flores. *Understanding Computers and Cognition: A New Foundation for Design*. Norwood, NJ: Ablex, 1986.
Wittgenstein, Ludwig. *Tractatus Logico-Philosophicus*. Translated by C. K. Ogden. Mineola, NY: Dover, 1999.

Index

Note: Page numbers with a *t* indicate tables.

abstraction. *See* levels of abstraction
actor-network theory (ANT), 23–25, 112–13, 115, 120–21; Harman's critique of, 126
addiction, 64–65, 67, 70, 91
aesthetics, xxx–xxxi, 79, 92–94, 139, 169–72
Alexander, Jennifer Karns, 154, 155, 157
"algorithmization," xxvii–xxviii, 133–34, 143
algorithms, vii, xvii, 99, 101, 141, 174; quasi-autonomous, xxvii–xxviii, 133–35, 143, 162; stock market, 134
All Watched Over by Machines of Loving Grace (documentary), 147–48
"allness" terms, 54–55
alterity relations, 22
Ames, Adelbert, Jr., 45
analog versus digital, 49–50, 55t, 57, 79, 83, 169, 171
Anders, Günther, 142
Angus, Ian, 13, 197n58
Anthropocene, viii, 59
"antimaterialist superstition," 8, 164
Anton, Corey, 12, 49, 90

Aristotle, 75, 120, 171; on causation, 60–62, 154; logic of, 53, 55; ontology of, 43
artificial intelligence (AI), xxiii, 133–34, 199n102
artificial life, 11
Augustine of Hippo, 4
autopoiesis theory, xxxvi, 71–74, 78; definition of, 72; reflexivity and, 11

background relations, 22
Bacon, Francis, 76
Bakardjieva, Maria, 150
Barad, Karen, 146, 147
Bateson, Gregory, 9–15; *Angels Fear*, 14, 47, 63, 129; career of, xxiv–xxv, 4–6; legacy of, 10–13; Mansell on, 148–49; *Mind and Nature*, xiii, 14, 47, 85; on schizophrenia, 3, 9, 12, 88–89, 99–100, 132, 148–49; *Steps to an Ecology of Mind*, xiii, 4, 63
Bateson, Mary Catherine, xxv, xxx, 69–71, 81, 83; on aesthetics, 93–94, 169; *Angels Fear*, 14, 47, 63, 129; on consciousness, 65; on epistemology, 46; *Our Own Metaphor*, 85–87, 186n56; on Pleroma, 49, 57, 124, 165

Bateson, Nora, xiii, xxv, xxx, 195n102
Bateson, William, 4
Beer, Stafford, 74
Bell, Daniel, 140, 156
Berman, Morris, xxxvi, 71, 84–85, 140, 168; on consciousness, 61, 75–76; on disenchantment, 61, 74–76; on efficiency, 154, 165; on purposiveness, 163
Bigelow, Julian, 62
Bijker, Wiebe, 24, 113
black box, 24–25, 27, 112–13, 115
Blake, William, 90
Boggs, Carl, 152
Borgmann, Albert, 105–7, 128
Bowers, Chet, 12
brain implants, xvii, xxiii, 101, 115
Brand, Stewart, 82
Brautigan, Richard, 147–48
"breakdown," Heidegger's notion of, 21, 37–38, 101, 109–10, 119, 125–29, 131
Bryant, Levi R., 193n80, 194n88
Bryngelsson, Erik, 170

capitalism, 67; Welzer on, 142, 152–53. *See also* neoliberalism
Capra, Fritjof, xxv, 11
central dichotomy, xxi, 17–20, 33–39, 47, 77–78; actor-network theory and, 23–25; critical theory of technology and, 25–27, 113; efficiency and, 144, 149–50, 154; Harman and, 123–24, 128–29; instrumentalism and, 104; postphenomenology and, 20–23. *See also* mind-matter problem
Charlton, Noel, 66, 80
Clarke, Bruce, 179n7
climate crisis, 32, 35, 83, 92
Coeckelbergh, Mark, xxx, 28, 136, 146

conscious purpose, xxx, 10, 59–68, 79–80, 102–5, 171; definition of, 64; instrumental reason and, 77; May on, 68; paradox of, 67, 80–87, 173; technology and, 68–71. *See also* purpose
consciousness, 10, 63; Berman on, 61, 75–77; Descartes on, 62; limits of, 65; "nonparticipating," 140; "participating," 75–77; purpose and, 60–64. *See also* unconscious
constructivism, xxvii, 150–51
Creatura. *See* Pleroma versus Creatura
Curtis, Adam, 147–48
"cybernetic ecology," 13, 148
cybernetic system, 66, 77
cybernetics, 48, 62, 163; history of, 6, 10–11; Macy Conferences on, 3, 6, 11
cyborgs, xvii, 105, 146–47, 174

Darby, Abraham, 140
Darwin, Charles, 155
de Botton, Alain, xxxii
Deacon, Terrence, 73–74
Deleuze, Gilles, xxv, 177n9
Descartes, René, 8–9, 105, 112, 124; Berman and, 75–76, 165; on consciousness, 62; Heidegger and, 182n35; Husserl and, 19
"designer's fallacy," 111
determinism, xix–xx, 85, 100, 104
Dewey, John, 111
digital, the, 166–67; analog versus, 49–50, 55t, 57, 79, 83, 169, 171
digital technologies, 167; "object" status of, 178n24; Rushkoff on, 138–39, 141
discourse, xxx, 32, 114, 143–49; efficiency and, 149–57, 161; matter and, 133, 143–49
"double bind," 3, 9, 12, 88–89, 148–49

Drucker, Peter, 152
Dubos, René, 84
Dutch East India Company, 124

ecological perspective, 28
ecology of mind, xxvi, xxxi, 3–4, 63, 107–8, 169
efficiency, xxi, 34, 100, 103, 129, 133; biological, 155; "cult" of, 153; definition of, 151–52; discourse and, 149–57, 161; Drucker on, 152; effectiveness versus, 151–52; "everyday," 34–35, 156; purpose of, 154, 157–61; reclaiming of, 157–61; of technology, 35–36, 149–57
Eliot, T. S., 139
Ellul, Jacques, xxvii, 34, 95, 150, 153, 157–58
embodiment relations, 21–22, 147
empirical turn, xxvii, xxviii, 28, 144–45
empiricism, 114, 168
epistemes, 17
epistemology, 7–10, 14, 76; "double description" of, 55t, 56–57; Epistemology versus, 45, 46, 51, 56, 124; ontology and, 43–49
essentialism, 85, 157–58; instrumentalism and, 100, 101, 182n35
extended mind thesis, 70

family therapy, 3, 6
Feenberg, Andrew, 25–28, 31, 113–16, 141, 195n105; on efficiency, 150–51; on Marcuse, xxxiv, 151; on "technical micropolitics," 27; technological invisibility and, 118; on "technological unconscious," 26, 113
figure-ground distinction, 29–30, 122–23

Flemons, Douglas, 13, 81, 82, 166
Floridi, Luciano, 136–37, 141
Fordism, 152
Foucault, Michel, viii, xxxi, 17
Frankfurt School, 25, 151, 152
Freud, Sigmund, 10, 80–81, 86, 89
Frisch, Max, 95

Gadamer, Hans-Georg, 61
Galileo, 110
general semantics, 12, 53, 184n36
genetically modified organisms (GMOs), xxiii
goals. *See* purposes
Guattari, Félix, xxv, 177n9

Habermas, Jürgen, 27, 106, 109
Halpern, Orit, 13
Haraway, Donna, 146
Harman, Graham, xxxvi, 21, 29–30, 57, 83; *Bells and Whistles*, 122–23; Hayler and, 194n98; Heidegger and, 120, 122; *Immaterialism*, 120, 124; on Latour, 25, 30, 194n88; on object-object interactions, 123, 127; object-oriented ontology of, 102, 113, 119–29, 144, 162–63, 169–70; as philosopher of technology, 193n74
Harries-Jones, Peter, 86
Hayler, Matt, 126, 129, 193n74, 194n98
Hayles, N. Katherine, 10–11, 146–47
health care, 52, 99, 134–35, 153
Heidegger, Martin, xxxi, 19–22, 27–28, 126; *Being and Time*, 19, 23; Borgmann and, 105; "breakdown" notion of, 21, 37–38, 101, 109–10, 119, 125–29; Ellul and, xxvii; on "Enframing," 18–19; Habermas and, 27, 109; Harman and, 120, 122; hermeneutic circle

Heidegger, Martin *(continued)* of, 180n37; Ihde on, 23, 182n35; Marcuse and, 195n105; ontology of, 122, 177n8, 182n35; on "the question concerning technology," xxv, 23; on "relational totality," 83, 110. *See also* Vorhandenheit versus Zuhandenheit
hermeneutic circle, 180n37
hermeneutic relations, 22
Hickman, Larry, 111
holism, 47, 63
"homeokinesis," 84
homeostasis, 11, 84
Horkheimer, Max, 150
Husserl, Edmund, 19–21, 110, 116, 170
hypomnesis, 108

Ihde, Don, 20, 38; on "designer's fallacy," 111; on Harman, 193n74; on Heidegger, 20–21, 23, 182n35; on Husserl, 20–21, 110, 116; linguistic turn and, 144; on "relational totality," 110–11; on visualization technologies, 22
immanence versus transcendence, 61
information and communication technology (ICT), xxvii–xxviii, 7–8, 135–39, 143; efficiency and, 155; Mansell on, 148
instrumentalism, xix–xxi, 18–19, 28, 99–104, 173; empirical turn and, xxviii; essentialism and, 85, 101, 182n35; postphenomenology and, 118
instrumentalization, 25–27
intellectual property rights, 148
Internet of Things (IoT), xxviii, 101, 134
"intra-action," 146
Islamic occasionalism, 121

Johnson, J. David, 161
Jongerden, Joost, 115–16

Jung, Carl Gustav, 47–48

Kant, Immanuel, 48, 73, 123, 127
Kauffman, Stuart, 73
Kelly, Kevin, 84
Kokoschka, Oskar, 76
Korzybski, Alfred, 53–56, 87, 90–91, 140; general semantics of, 12, 53, 184n36; on "time-binding," 12, 53, 184n39

Latour, Bruno, 9, 31, 106, 142, 181n13; actor-network theory of, 23–25, 112–13, 115, 120–21; *Facing Gaia*, 192n43; Harman on, 25, 30, 194n88; *An Inquiry into Modes of Existence*, 192n43; new materialism and, 120–21, 193n79; Pickering and, 144; semiotics and, 145
Lemmens, Pieter, 115–16
levels of abstraction, 12, 51, 87–95, 104, 118–19; disappearing technologies and, 133–41; Floridi on, 137–38, 141; Mansell on, 148–49; schizophrenia and, 132; technological instrumentation and, 110–11
linguistic turn, 144–45
Locke, John, 62
"logical types." *See* levels of abstraction
Luhmann, Niklas, 74, 194n88

Macy Conferences on cybernetics, 3, 6, 11
Maggini, Golfo, 193n74
"managementalization," 52
Mansell, Robin, 13, 147–49, 168
map versus territory, 12, 52–57, 55t, 79, 140
Marchand, Philip, 178n20
Marcuse, Herbert, 28, 113–14, 117, 138; on abstraction, 140–41; Feenberg and, xxxiv, 151

material turn, xxix–xxx, 31, 144–45, 149
materialism, 8, 125, 164; idealism versus, 8, 44; new, 120–21, 193n79. *See also* mind-matter problem
"materialist superstition," 8, 164
Maturana, Humberto R., xxxvi, 11, 71–77, 84, 173–74, 187n76
May, Rollo, 5, 46, 68, 84
Mazlish, Bruce, xxix
McCulloch, Warren, 11
"McDonaldization," 152
McGinn, Robert E., 102, 103
McLuhan, Marshall, xii, 37, 61; on figure-ground distinction, 29–30, 122–23; media ecology and, 12, 29, 30t; on "somnambulists," 178n20; on technological advances, 69; on technology as body's extension, 70, 102
Mead, Margaret, 3, 5, 84
media ecology, 11–12, 29, 30t
metacommunication, 168
"metalogues," 112, 190n7
mind, xxx, 47, 56, 59–60, 62–63, 70–71; ecology of, xxvi, xxxi, 3–4, 63, 107–8, 169; information and, 7–8; systems of knowledge and, 69–71
mind-matter problem, xxx, 59, 100, 105, 112, 124, 125; Berman on, 75–76, 165; epistemology of, 7–10. *See also* central dichotomy
Mitcham, Carl, 151
Montaigne, Michel de, xxxi
"moral imagination," 142–43
Morozov, Evgeny, 83–85, 155
Morton, Timothy, 195n105
Mumford, Lewis, 157

neoliberalism, vii, 67, 117, 147–48; efficiency of, 152–53, 165
neural implants, xvii, xxiii, 101, 115

new age philosophy, 8, 44, 164, 172
new materialism, 120–21, 193n79
Nietzsche, Friedrich, xi–xxi

Oakeshott, Michael, 86
object-oriented ontology (OOO), 102, 113, 119–29, 144, 162–63, 169–70
occasionalism, Islamic, 121
Oceanic Institute (Hawaii), 6
Ockham, William of, 154
ontology, 43–49; Aristotle's, 43; Harman's, 102, 113, 119–29, 144, 162–63, 169–70; Heidegger's, 122, 177n8, 182n35; Plato's, 43
Oosterling, Henk, 12
operationalism, 114, 117
"originary technicity," 109
ouroboros, x

panpsychism, 7, 75
pattern. *See* quality versus quantity
Pavlov, Ivan, 88
pharmakon, 107
phenomenology, 19, 43, 46, 115. *See also* postphenomenology
Pickering, Andrew, 144
Pinch, Trevor, 24, 113
Plato, 14, 55, 61, 139, 154; ontology of, 43
play, 12, 61, 90, 93, 94, 110, 199n99
Pleroma versus Creatura, 9, 56–59, 79, 164–65; definitions of, 47–49, 55t; Harman and, 123–24, 128
posthumanism, viii, 10–11, 146–47
postmodernism, viii, 23–24, 174
postphenomenology, xxvii, 20–23, 38, 77, 110–11; instrumentalism and, 118; material turn in, 144–45; scholars of, 115–16. *See also* phenomenology
poststructuralism, viii, 43
pragmatism, 111

presence-at-hand. *See* Vorhandenheit versus Zuhandenheit
psyche, 63–64, 125, 194n95
psychedelics, 94
Puech, Michel, 31–32, 139–40, 161, 196n43
purification-proliferation dichotomy, 25
purposes, 35, 60–64, 128, 172–74; of efficiency, 154, 157–61; goals and, 59–60, 171; of technology, xxxiv, 160–61, 163; teleology and, 163. *See also* conscious purpose
purposive structures, 103–4, 127, 136, 171–73

quality versus quantity, xxvii–xxviii, 8, 50–52, 55t, 169, 171
quantification, 50–52, 165–66
quantum physics, 53

Rao, Mithun Bantwal, 115–16
rationalization, "subversive," 25–26, 113, 115
readiness-to-hand. *See* Zuhandenheit versus Vorhandenheit
realism: "agential," 146; "instrumental," 110; speculative, 123
"relational totality," 83, 110
relationalism, 21, 43–47, 56–58, 110; Barad on, 146–47; Harman on, 120–21; Latour and, 113
relativity theory, 53
Ritzer, George, 152
romanticism, 145
Rorty, Richard, 86
Rosenblueth, Arturo, 62
Ross, Daniel, 107–8
Ruivenkamp, Guido, 116
Rushkoff, Douglas, 138–39, 141
Russell, Bertrand, 88, 136

Sanders, J. W., 136–37

Santiago Theory of Cognition, 71
Sartre, Jean-Paul, xxxi
Savino, Gianfranco, 180n35
"schismogenesis," xxv
schizophrenia, xxi, 36, 93; Bateson's theory of, 3, 9, 12, 88–89, 99–100, 132, 148–49; paranoia and, 136, 138
Schönberg, Arnold, 93
Schwartz, Theodore, 186n56
science and technology studies (STS), 5, 13, 116
search engines, 138–39, 143
semantics, general, 12, 53, 184n36
Serres, Michel, 139, 141, 196n23
Sharon, Tamar, 28
Shirky, Clay, 84
Simondon, Gilbert, 108
Smith, Dominic, xxviii, 116, 145–46, 168, 196n41, 196n42
Snow, C. P., 61
social construction of technology (SCOT), 26, 145, 150–51
social media, 135, 160
"solutionism," 83–85
Son, Wha-Chul, 34, 151, 158, 159
Stehr, Nico, 140
Stein, Janice Gross, 153–55, 157, 159, 165
Stephenson, George, 140
Stiegler, Bernard, xxv, 107–9, 127
Strate, Lance, 94, 95
substantivism, 44
"subversive rationalization," 25–26, 113, 115
Swaab, Dick, 44
symbolic language, 94–95, 144, 161
"systemic wisdom," 66–67, 70, 78, 80–81, 89–90, 118–19, 135–36, 173, 188n3
systems theory, 103, 148, 163, 199n102

Taoism, 12–13, 81, 117, 198n98
tautology, xix, 57, 185n54
Taylor, Frederick, 155
"technical micropolitics," 27, 113
"technological unconscious," 26, 31, 113, 196n43
technology, 75–76; conscious purpose and, 68–71, 102–5; critical theory of, xxvii, 25–27, 113–19, 150; definitions of, 18, 20, 35, 70, 84, 102; "disappearing," 99–101, 105–7, 118–19, 131, 133–43; efficiency and, 35–36, 149–57; "excesses" of, 106; Habermas's definition of, 27; Hayler's definition of, 126; Hickman's definition of, 111; living with, xi–xii, xxx–xxxiii, 129, 131–33, 178n22; philosophy of, xix–xxix, 17–20, 99–102, 105–19, 125–28; purpose of, x, xxxvi, 160–61, 163; Stiegler's definition of, 109
teleology, 62, 106, 163
territory. *See* map versus territory
Teschner, George, 198n98
thermostats, 48
Thompson, Evan, 63
Tomasi, Alessandro, 198n98
tool analysis. *See* Vorhandenheit versus Zuhandenheit
Trump, Donald, vii, 92

unconscious, 45–46, 67, 80–81, 169; collective, 100; "technological," 26, 31, 113, 196n43. *See also* consciousness

van Boeckel, Jan, 12, 93
Varela, Francisco J., xxxvi, 11, 63, 71–77, 84, 173–74, 183n11, 187n76
Verbeek, Peter-Paul, xxx, 38–39, 115–17, 120, 174; on Borgmann, 106; on material turn, 144–45; *Moralizing Technology,* 29
virtuality, 11, 147
visualization technologies, 22
Vorhandenheit versus Zuhandenheit, 19–22, 30t, 37–38, 86, 119–20, 131–32

Wark, McKenzie, 141
Warren, Rick, 160
Watt, James, 140
Weber, Andreas, 73
Weber, Max, 25, 27, 75, 77
Webster, Frank, 140, 156
Welzer, Harald, 142–43, 152–53, 157
Whitehead, Alfred North, 43, 75, 88
Wiener, Norbert, 62
Wikipedia, 143
Wilden, Anthony, 148, 166–68
Wiltse, Heather, 101, 127
Winner, Langdon, 159, 198n92
Wittgenstein, Ludwig, 14

Zen Buddhism, 33
Zuckerberg, Mark, 160
Zuhandenheit versus Vorhandenheit, 19–22, 30t, 37–38, 86, 119–20, 131–32

www.ingramcontent.com/pod-product-compliance
Lightning Source LLC
Chambersburg PA
CBHW030537230426
43665CB00010B/932